Brouwer's Cambridge lectures on intuitionism

Brouwer's Cambridge lectures on intuitionism

Edited by

D. VAN DALEN
Rijksuniversiteit Utrecht

CAMBRIDGE UNIVERSITY PRESS
CAMBRIDGE
LONDON NEW YORK NEW ROCHELLE
MELBOURNE SYDNEY

CAMBRIDGE UNIVERSITY PRESS
Cambridge, New York, Melbourne, Madrid, Cape Town, Singapore,
São Paulo, Delhi, Dubai, Tokyo, Mexico City

Cambridge University Press
The Edinburgh Building, Cambridge CB2 8RU, UK

Published in the United States of America by Cambridge University Press, New York

www.cambridge.org
Information on this title: www.cambridge.org/9780521177368

First published 1981
First paperback edition 2011

A catalogue record for this publication is available from the British Library

ISBN 978-0-521-23441-2 Hardback
ISBN 978-0-521-17736-8 Paperback

Contents

Editorial preface

In the years after the Second World War, from 1946 to 1951, Brouwer gave various series of lectures on intuitionism at the University of Cambridge. He decided to collect the material together in a monograph, which would be the first systematic exposition of intuitionistic mathematics in book form. As a matter of fact it was not his first such enterprise; in a way Brouwer meant his dissertation to be a comprehensive exposition of his philosophical and mathematical views. In 'The rejected parts of Brouwer's dissertation on the foundations of mathematics',† van Stigt tells how the intervention of Brouwer's thesis adviser prevented the insertion of the philosophical parts. Although he never got beyond the stage of intentions, Brouwer seriously considered the publication of a revised version of his thesis. In 1929 the Noordhoff publishing company invited him to carry out these plans, but Brouwer was at that time already involved in the preparation of a German monograph. From the late twenties onwards Brouwer gave expository and propagandistic lectures, mainly in Germany but also in other countries. According to eye-witnesses Brouwer, with his lectures, made quite an impression. The intuitionist 'putsch' was the topic of many a heated discussion.

In particular his *Gastvorlesungen* in Berlin in 1927 raised high hopes for the establishment of intuitionism as an accepted part of mathematical practice. The barely veiled rivalry between Berlin and Göttingen most probably played an important role in the enthusiastic reception of intuitionism in Berlin; the spirited

† van Stigt, W. P. (1979) The rejected parts of Brouwer's dissertation on the foundations of mathematics. *Historia Mathematica* 6, 385–404.

challenge of Hilbert's complacent supremacy was welcomed by more than one mathematician, especially in Berlin. The *Berliner Tageblatt* invited Brouwer to contribute a series of articles on intuitionism in a public debate with Hilbert. At the same time the publishing house of Walter de Gruyter asked Brouwer if it could publish his Berlin lectures. In March 1927, when a first draft of the text was ready, de Gruyter proposed to Brouwer that he should expand the text into a book 'which the public no doubt will tear out of your hands'. Unfortunately nothing came of it; Brouwer dropped the Berlin lectures and started to work on the book, but never finished it.

Apart from his course on intuitionism in Amsterdam (1927/28), Brouwer also lectured in Groningen (1933), Geneva (1934), Cape Town (1952), the USA and Canada (1953).

These courses were all modelled after the Berlin lectures. The Vienna guest lecture, 'Die Struktur des Kontinuums',† the Cape Town lectures,‡ and his Canadian and American lectures§ are in a sense selections of Brouwer's projected *magnum opus*.

Eventually none of the projected books or monographs appeared. When van Stigt and I undertook the collecting of Brouwer's estate in 1976 we found, through the kind assistance of Professor Dijkman, various manuscripts, among which was one of the Berlin lectures and one of the Cambridge lectures. The Berlin lectures had not been rewritten or greatly changed, but the Cambridge lectures showed signs of being extensively reworked for a longer period. The manuscript we found carried the handwritten note (in Brouwer's handwriting) *Laatste herziening tekst van Cambridge-boek* (Final revision text of Cambridge book), so there is some reason to believe that it represents Brouwer's views (of the fifties) in a reasonably adequate way.

It is an open question why Brouwer never published his

† Brouwer, L. E. J. (1930) *Die Struktur des Kontinuums*. Gistel, Vienna. 14 pp.

‡ Brouwer, L. E. J. (1952) Historical background, principles and methods of intuitionism. *South African Journal of Science* **49**, 139–46.

§ Brouwer, L. E. J. (1954) Points and spaces. *Canadian Journal of Mathematics* **6**, 1–17.

lectures. In 1951, when Brouwer lectured for the last time in Cambridge, five of the six planned chapters were finished (the sixth chapter would probably have dealt with the theory of functions). Professor S. W. P. Steen and Dr N. Routledge, who were involved in the preparation of the Cambridge lectures, had the impression that Brouwer never intended to publish the lectures, and that he used the proposed book as a pretext to return to Cambridge where he loved to lecture. Nevertheless, Brouwer kept revising his manuscript long after he stopped lecturing in Cambridge.

Comparing the Cambridge lectures with earlier expositions, e.g. the Berlin lectures, it will strike the reader that Brouwer did not strive for new revolutionary breakthroughs, but rather for a consolidation of the fundamental material of intuitionism. I think that it is safe to agree that Brouwer's two most spectacular performances in his foundational work, given the basic notions (e.g. natural number, choice sequence), are *the continuity theorem* (involving continuity, bar induction, etc.) and *the strong counter-examples* (involving the creative subject). Although the ultimate exploitation of the notion of the creative subject is achieved only in 1949,† the notion already occurs implicitly in *Die Struktur des Kontinuums* and in the Berlin lectures (1927). In the Cambridge lectures extensive use is made of the creative subject, but *not* in the optimal way (most counterexamples are weak, i.e. in the form 'as long as non-tested statements are known we cannot say that', and not strong, i.e. of the form '$\neg A$ holds' or, in Brouwer's terminology, 'A is absurd'). It remains a baffling puzzle why Brouwer did not fully exploit the strength of the creative subject. Possibly he arrived at the stronger results (what he calls the contradictority of, for example, the classical theory of functions) only after he had designed his course, so that he did not choose to embark on a large-scale overhaul of his lecture notes. It is

† Brouwer, L. E. J. (1949a) De non-aequivalentie van de constructieve en de negatieve orderelatie in het continuum. *Indagationes Mathematicae* **11**, 37–9. Translated as 'The non-equivalence of the constructive and negative order relation on the continuum' in Brouwer, L. E. J. (1975) *Collected works*, Vol. I, ed. A. Heyting. North-Holland Publ. Co., Amsterdam. pp. 495–6.

also conceivable that Brouwer had second thoughts on the matter. It is remarkable that he published his first strong application of the creative subject in 1949 in Dutch,† whereas the first such application in English appeared only in 1954.‡ Notes written in pencil in the margin of the manuscript indicate that Brouwer entertained the idea of strengthening the presentation. He never did so, however.

It will be convenient to keep the notion of the creative subject, e.g. in Kreisel's presentation,[§] in mind when reading the text. It will be helpful to sort out the weak from the strong negations. Posy has pointed out that, in particular where virtual and inextensible order are concerned, an analysis in terms of $\vdash_n$ is useful.[||]

On the one hand Brouwer tries to obtain maximal generality; in particular in the theory of spreads he adds a great number of distinctions which result from a generous generalization of the basic notions as presented in earlier expositions.[¶] On the other hand he wishes to restrict the notion of choice sequence by abolishing higher order restrictions (i.e. restrictions on restrictions, etc.), which were explicitly introduced in Brouwer (1942a).* In the Cambridge lectures Brouwer is vaguely suspicious of higher order restrictions: 'But at present the author is inclined to think this admission superfluous and perhaps leading to needless complications'. In Brouwer (1952)** this has been strengthened to: 'However, this admission is not justified by

† Brouwer, L. E. J. (1949a) De non-aequivalentie van de constructieve en de negatieve orderelatie in het continuum. *Indagationes Mathematicae* **11**, 37–9.

‡ Brouwer, L. E. J. (1954) An example of contradictority in classical theory of functions. *Indagationes Mathematicae* **16**, 204–5.

§ Kreisel, G. (1967) Informal rigour and completeness proofs. In *Problems in the philosophy of mathematics*, ed. I. Lakatos, pp. 138–86. North-Holland Publ. Co., Amsterdam.

|| Posy, C. J. A note on Brouwer's definitions of unextendable order (to appear).

¶ Brouwer, L. E. J. (1925) Zur Begründung der intuitionistischen Mathematik I. *Mathematische Annalen* **93**, 244–57.

* Brouwer, L. E. J. (1942a) Zum freien Werden von Mengen und Funktionen. *Indagationes Mathematicae* **4**, 107–8.

** Brouwer, L. E. J. (1952) Historical background, principles and methods of intuitionism. *South African Journal of Science* **49**, 139–46.

close introspection and moreover would endanger the simplicity and rigour of further developments'. Unfortunately Brouwer has not elaborated the point, and so far no convincing arguments have come forward to decide the issue for or against higher order restrictions. The matter is of some interest as it has been argued that, according to Brouwer (1952), lawless sequences are not intuitionistically acceptable. It is curious that Brouwer never mentions this simplest of all notions of choice sequence in print; he mentions it, however, explicitly in a letter to Heyting in 1924. Freudenthal recalls that when, during the Berlin lectures, the notion of a sequence based on throwing a die was suggested Brouwer firmly rejected it (probably because it was based on the physical world).

A feature that might irritate a modern reader, but which typically belongs to Brouwer, is the consistent refusal to use symbolic notation. It seems tempting to ascribe this to Brouwer's aversion for formalization with its Hilbertian undertones. I think, however, that it was simply a characteristic of Brouwer's style. Even in his topological papers he writes in a leisurely style, avoiding the economic use of handy formalisms. In the present monograph, for example, Brouwer consistently uses such expressions as 'the absurdity of the absurdity of α', 'α is contradictory', where $\neg\neg\alpha$ and $\neg\alpha$ would be infinitely more readable. Still in 1951 Brouwer uses the old terminology of Schoenfliess for union and intersection, $\mathfrak{P}(M,N)$ and $\mathfrak{D}(M,N)$, but even in this instance Brouwer does not make an effective use of the notation. Almost always he prefers 'the union of M and N' to '$\mathfrak{P}(M,N)$'.

Taking into account Brouwer's views on communication it seems more reasonable to state that his style was a logical consequence of his personal experience and technique of transferring knowledge and insight. That is to say, in particular in the case of intuitionism that cannot be taught as if it were, say, linear algebra, there was a strong aspect of convincing, conversion, especially in the eyes of Brouwer who considered intuitionistic mathematics as the one and only correct mathematics. As a result a persuasive, personal, non-formal style is exactly what one would expect. The only change in notation I have made is to use the standard union and intersection symbols.

Together with the manuscript a number of scraps and private notes for use during the lectures were filed. I have collected some of them into an appendix, which also contains parts of an address he gave in 1951. Furthermore I have added a number of notes of Brouwer that are relevant to the matter and that may be clarifying. Notes by the editor in the text are contained in square brackets [ ]. In addition to Brouwer's own footnotes, indicated by symbols, there are numbered endnotes provided by the editor.

Thanks are due to Professor Dijkman, who donated all the material to the Brouwer Archive, and also to the Mathematics Department of the Rijksuniversiteit Utrecht for its generous secretarial and material assistance.

I am particularly indebted to W. P. van Stigt, who with great enthusiasm and ingenuity joined in the researches connected with Brouwer's estate, and also to the Netherlands Organization for the Advancement of Pure Research (Z.W.O.) that enabled van Stigt to spend the academic year 1976/77 in Utrecht.

In the matter of the collecting of the biographical and scientific material the Brouwer family, in the person of the executor of the will, Ir. L. E. J. Brouwer, has been very helpful. I am grateful for his generous cooperation in the founding of the Brouwer Archive and for his approval of the publication of the material.

I want to express my gratitude to the Cambridge University Press for its willingness to legitimize the present, rather aged, child and for the helpfulness of its staff, in particular Mrs Jane Holland, in the preparation of the manuscript.

D. van Dalen Utrecht, June 1980

1

Historical introduction and fundamental notions[1]

The gradual transformation of the mechanism of mathematical thought is a consequence of the modifications which, in the course of history, have come about in the prevailing philosophical ideas, firstly concerning the origin of mathematical certainty, secondly concerning the delimitation of the object of mathematical science. In this respect we can remark that in spite of the continual trend from object to subject of the place ascribed by philosophers to time and space in the subject–object medium, the belief in the existence of immutable properties of time and space, properties independent of experience and of language, remained well-nigh intact far into the nineteenth century. To obtain exact knowledge of these properties, called mathematics, the following means were usually tried: some very familiar regularities of outer or inner experience of time and space were postulated to be invariable, either exactly, or at any rate with any attainable degree of approximation. They were called axioms and put into language. Thereupon systems of more complicated properties were developed from the linguistic substratum of the axioms by means of reasoning guided by experience, but linguistically following and using the principles of classical logic. We will call the standpoint governing this mode of thinking and working the *observational* standpoint, and the long period characterized by this standpoint the observational period. It considered logic as autonomous, and mathematics as (if not existentially, yet functionally) dependent on logic.

For space the observational standpoint became untenable when, in the course of the nineteenth and the beginning of the twentieth centuries, at the hand of a series of discoveries with which the names of Lobatchefsky, Bolyai, Riemann, Cayley, Klein, Hilbert, Einstein, Levi-Cività and Hahn are associated, mathematics was gradually transformed into a mere science of numbers; and when besides observational space a great number of other spaces, sometimes exclusively originating from logical speculations, with properties distinct from the traditional, but no less beautiful, had found their arithmetical realization. Consequently the science of classical (Euclidean, three-dimensional) space had to continue its existence as a chapter without priority, on the one hand of the aforesaid (exact) science of numbers, on the other hand (as applied mathematics) of (naturally approximative) descriptive natural science.

In this process of extending the domain of geometry, an important part had been played by the *logico-linguistic method*, which operated on words by means of logical rules, sometimes without any guidance from experience and sometimes even starting from axioms framed independently of experience. Encouraged by this the *Old Formalist School* (Dedekind, Cantor, Peano, Hilbert, Russell, Zermelo, Couturat), for the purpose of a rigorous treatment of mathematics *and logic* (though not for the purpose of furnishing objects of investigation to these sciences), finally rejected any elements extraneous to language, thus divesting logic and mathematics of their essential difference in character, as well as of their autonomy. However, the hope originally fostered by this school that mathematical science erected according to these principles would be crowned one day with a proof of its non-contradictority was never fulfilled, and nowadays, after the logical investigations performed in the last few decades, we may assume that this hope has been relinquished universally.

Of a totally different orientation was the *Pre-intuitionist School*, mainly led by Poincaré, Borel and Lebesgue. These thinkers seem to have maintained a modified observational standpoint for the introduction of natural numbers, for the principle of complete induction, and for all mathematical entities springing from this source without the intervention of axioms of existence, hence for what might be called the 'separable' parts of arithmetic and of algebra. For these, even for such theorems as were deduced by means of classical logic, they postulated an existence and exactness independent of language and logic and regarded its non-contradictority as certain, even without logical proof. For the continuum, however, they seem not to have sought an origin strictly extraneous to language and logic. On some occasions they seem to have contented themselves with an ever-unfinished and ever-denumerable species of 'real numbers' generated by an ever-unfinished and ever-denumerable species of laws defining convergent infinite sequences of rational numbers. However, such an ever-unfinished and ever-denumerable species of 'real numbers' is incapable of fulfilling the mathematical function of the continuum for the simple reason that it cannot have a positive measure. On other occasions they seem to have introduced the continuum by having recourse to some logical axiom of existence, such as the 'axiom of ordinal connectedness', or the 'axiom of completeness', without either sensory or epistemological evidence. In both cases in their further development of mathematics they continued to apply classical logic, including the *principium tertii exclusi*, without reserve and independently of experience. This was done regardless of the fact that the non-contradictority of systems thus constructed had become doubtful by the discovery of the well-known logico-mathematical antonomies.

In point of fact, pre-intuitionism seems to have maintained on the one hand the essential difference in character between logic and mathematics, and on the other hand the autonomy of logic, and of a part of mathematics. The rest of mathematics became dependent on these two.

Meanwhile, under the pressure of well-founded criticism exerted upon old formalism, Hilbert founded the *New Formalist School*, which postulated existence and exactness independent of language not for proper mathematics but for meta-mathematics, which is the scientific consideration of the symbols occurring in perfected mathematical language, and of the rules of manipulation of these symbols. On this basis new formalism, in contrast to old formalism, *in confesso* made primordial practical use of the intuition of natural numbers and of complete induction. It is true that only for a small part of mathematics (much smaller than in pre-intuitionism) was autonomy postulated in this way. New formalism was not deterred from its procedure by the objection that between the perfection of mathematical language and the perfection of mathematics itself no clear connection could be seen.

So the situation left by formalism and pre-intuitionism can be summarized as follows: for the elementary theory of natural numbers, the principle of complete induction and more or less considerable parts of arithmetic and of algebra, exact existence, absolute reliability and non-contradictority were universally acknowledged, independently of language and without proof. As for the continuum, the question of its languageless existence was neglected, its establishment as a set of real numbers with positive measure was attempted by logical means and no proof of its non-contradictory existence appeared. For the whole of mathematics the four principles of classical logic were accepted as means of deducing exact truths.

In this situation intuitionism intervened with two acts, of which the first seems to lead to destructive and sterilizing consequences, but then the second yields ample possibilities for new developments.

FIRST ACT OF INTUITIONISM *Completely separating mathematics from mathematical language and hence from the phenomena of language described by theoretical logic, recognizing that intuitionistic mathematics is an essentially languageless activity of the mind having its origin in the perception of a move of time. This perception of a move of time may be described as the falling apart of a life moment into two distinct things, one of which gives way to the other, but is retained by memory. If the twoity thus born is divested of all quality, it passes into the empty form of the common substratum of all twoities. And it is this common substratum, this empty form, which is the basic intuition of mathematics.*

Inner experience reveals how, by unlimited unfolding of the basic intuition, much of 'separable' mathematics can be rebuilt in a suitably modified form. In the edifice of mathematical thought thus erected, language plays no part other than that of an efficient, but never infallible or exact, technique for memorizing mathematical constructions, and for communicating them to others, so that mathematical language by itself can never create new mathematical systems. But because of the highly logical character of this mathematical language the following question naturally presents itself. *Suppose that, in mathematical language, trying to deal with an intuitionist mathematical operation, the figure of an application of one of the principles of classical logic is, for once, blindly formulated. Does this figure of language then accompany an actual languageless mathematical procedure in the actual mathematical system concerned?*

A careful examination reveals that, briefly expressed, the answer is *in the affirmative, as far as the principles of contradiction and syllogism are concerned,*[2] *if one allows for the inevitable inadequacy of language as a mode of description and communication. But with regard to the principle of the excluded third, except in special cases, the answer is in the negative, so that this principle cannot in general serve as an instrument for discovering new mathematical truths.*

Indeed, if each application of the *principium tertii exclusi* in mathematics accompanied some actual mathematical procedure, this would mean that each mathematical assertion (i.e. an assignment of a property to a mathematical entity) could be *judged*, that is to say could either be proved or be reduced to absurdity.

Now every construction of a bounded finite nature in a finite mathematical system can only be attempted in a finite number of ways, and each attempt proves to be successful or abortive in a finite number of steps. We conclude that every assertion of possibility of a construction of a bounded finite nature in a finite mathematical system can be judged, so that in these circumstances applications of the *principium tertii exclusi* are legitimate.

But now let us pass to infinite systems and ask for instance if there exists a natural number n such that in the decimal expansion of π the nth, $(n+1)$th, ..., $(n+8)$th and $(n+9)$th digits form a sequence 0123456789. This question, relating as it does to a so far not judgeable assertion, can be answered neither affirmatively nor negatively. But then, from the intuitionist point of view, because outside human thought there *are* no mathematical truths, the assertion that in the decimal expansion of π a sequence 0123456789 either does or does not occur is devoid of sense.

The aforesaid property, suppositionally assigned to the number n, is an example of a *fleeing property*,[3] by which we understand a property f, which satisfies the following three requirements:

(i) for each natural number n it can be decided whether or not n possesses the property f;
(ii) no way of calculating a natural number n possessing f is known;
(iii) the assumption that at least one natural number possesses f is not known to be an absurdity.

Obviously the fleeing nature of a property is not necessarily permanent, for a natural number possessing f might at some time be found, or the absurdity of the existence of such a natural number might at some time be proved.

By the *critical number* κ_f of the fleeing property f we understand the (hypothetical) smallest natural number possessing f.[4] A natural number will be called an *up-number* of f if it is not smaller than κ_f, and a *down-number* if it is smaller than κ_f. Of course, f would cease to be fleeing if an up-number of f were found.

A fleeing property is called *two-sided with regard to parity* if neither of an odd nor of an even κ_f the absurdity of existence has been demonstrated. Let s_f be the real number which is the limit of the infinite sequence $a_1, a_2, \ldots$, where $a_\nu = (-2)^{-\nu}$ if ν is a down-number and $a_\nu = (-2)^{-\kappa_f}$ if ν is an up-number of f. This real number violates the principle of the excluded third, for neither is it equal to zero nor is it different from zero and, although its irrationality is absurd, it is not a rational number. Moreover if f is two-sided with regard to parity then s_f is neither $\geqq 0$ nor $\leqq 0$.

The belief in the universal validity of the principle of the excluded third in mathematics is considered by the intuitionists as a phenomenon of the history of civilization of the same kind as the former belief in the rationality of π, or in the rotation of the firmament about the earth. The intuitionist tries to explain the long duration of the reign of this dogma by two facts: firstly that within an arbitrarily given domain of mathematical entities the non-contradictority of the principle for a single assertion is easily recognized; secondly that in studying an extensive group of simple every-day phenomena of the exterior world, careful application of the whole of classical logic was never found to lead to error.†

The mathematical activity made possible by the first act of intuitionism seems at first sight, because mathematical creation by means of logical axioms is rejected, to be confined to 'separable' mathematics, mentioned above; while, because also the principle of the excluded third is rejected, it would seem that even within 'separable' mathematics the field of activity would have to be considerably curtailed. In particular, since the continuum appears to remain outside its scope, one might fear at this stage that in intuitionism there would be no place for analysis. But this fear would have assumed that infinite se-

† This means *de facto* that common objects and mechanisms subjected to familiar manipulations behave as if the system of states they can assume formed part of a finite discrete set, whose elements are connected by a finite number of relations.

quences generated by the intuitionistic unfolding of the basic intuition would have to be fundamental sequences,[5] i.e. predeterminate infinite sequences proceeding, like classical ones, in such a way that from the beginning the nth term is fixed for each n. Such however is not the case; on the contrary, a much wider field of development, including analysis and often exceeding the frontiers of classical mathematics, is opened by the second act of intuitionism.

SECOND ACT OF INTUITIONISM *Admitting two ways of creating new mathematical entities: firstly in the shape of more or less freely proceeding infinite sequences of mathematical entities previously acquired* (so that, for example, infinite decimal fractions having neither exact values[6], nor any guarantee of ever getting exact values are admitted); *secondly in the shape of mathematical species, i.e. properties supposable for mathematical entities previously acquired, satisfying the condition that if they hold for a certain mathematical entity, they also hold for all mathematical entities which have been defined to be 'equal' to it, definitions of equality having to satisfy the conditions of symmetry, reflexivity and transitivity.*

By the *elements* of a species we understand the mathematical entities previously acquired† for which the property in question holds.‡ These elements will also be said to *belong* to the species.

Two mathematical entities will be called *different* if their equality proves to be absurd. The notations for equality and difference will be $=$ and $\neq$ respectively.

Two infinite sequences of mathematical entities $a_1, a_2, \ldots$ and $b_1, b_2, \ldots$ will be said to be *equal*, or *identical*, if $a_\nu = b_\nu$ for each ν and *distinct* if a natural number n can be indicated (or calculated) such that a_n and b_n are different.

† It follows that during the development of intuitionist mathematics some species may have to be considered as being tacitly defined again and again in the same way.

‡ A species can be an element of another species, but never an element of itself!

A species is called *discrete* if any two of its elements can be proved either to be equal or to be different.

If the species M possesses an element which cannot possibly belong to the species N, or, what is the same, is different from each element of N, we shall say that M *deviates* from N.

The species M will be called a *subspecies* of the species N, and we shall write $M \subset N$, if every element of M can be proved to belong to N. If in addition N deviates from M, then M is called a *proper subspecies* of N. If each element of N either belongs to M or cannot possibly belong to M, then M is called a *removable subspecies* of N.

Two species are said to be *equal*, or *identical*, if for each element of either of them an element of the other equal to it can be indicated. They are called *different* if their equality is absurd, and *congruent* if neither can deviate from the other.[7] For instance the following three species are congruent to one another: the species P of the infinite sequences of 0s and 1s; the species M of those elements of P which either consist only of 0s or begin with a 1 or begin with a finite number of 0s followed by a 1; and the species N of those elements of P which either consist of 1s or begin with a 0, or begin with a finite number of 1s followed by a 0.

Obviously each property that is absurd for all elements of one of two congruent species is also absurd for those of the other. Suppose, conversely, that every property that is absurd for the elements of the one species is also absurd for the elements of the other. Then this holds in particular for the property of being different from each element of one of the two species, from which we immediately deduce the congruence of the two species. Note that two species can be congruent and different at the same time.

Let S be a species of species s. The property of being an element of all these species s will be called the *intersection* $\cap S$,[8] and the property of being an element of at least one of the species s will be called the *union* $\cup S$ of the species s.[9] If S has only a finite number of elements $s_1, \ldots, s_n$, or a fundamental sequence $s_1, s_2, s_3, \ldots$, we shall also write $s_1 \cap s_2 \cap \ldots \cap s_n, s_1 \cup s_2 \cup \ldots \cup s_n$, and $\cap_{i \in N} s_i$, $\cup_{i \in N} s_i$ respectively.[10]

A species which cannot possess an element is said to be *empty*. Two different species whose intersection is empty are called *disjoint*.

The above-mentioned well-nigh evident non-contradictority within an arbitrarily given domain of mathematical entities of the *simple principle of judgeability* or the *simple principle of the excluded third*, i.e. of the principle of the excluded third enunciated for an arbitrary single assertion, holds for the whole of intuitionistic mathematics. It even holds for the simultaneous enunciation of the principle of the excluded third for an arbitrary finite number of assertions. For, the enunciation of the principle of the excluded third for an assertion is itself an assertion, and *finite additivity of non-contradictority* in an arbitrarily given domain of mathematical assertions is easily established in the following way.

Let the assertions ρ and σ be non-contradictory, and let us start for a moment from the supposition ω that the conjunction τ of ρ and σ is contradictory. Then the truth of ρ would entail the contradictority of σ. Since the contradictority of σ clashes with the data, the truth of ρ is absurd, i.e. ρ is absurd. Thus a consequence of the supposition ω clashes with the data, and so this supposition is contradictory, i.e. τ is non-contradictory.

This finite additivity of the non-contradictority of the principle of judgeability cannot be extended to universal additivity; in particular the contradictority can be proved of the following *complete principle of judgeability.*

If a, b and c are species of mathematical entities, and both a and b are subspecies of c, and b consists of those elements of c which cannot belong to a, then c is identical with the union of a and b.

We formulate the following pair of corollaries of the simple and the complete principles of judgeability respectively, of which the former is non-contradictory and the latter contradictory:

(1) if of two assertions *a* and *b*, *a* is equivalent to the absurdity of *b*, then *b* is equivalent to the absurdity of *a* (*simple principle of reciprocity of absurdity* or *simple principle of truth by non-contradiction*);

(2) if *a*, *b* and *c* are species of mathematical entities, and both *a* and *b* are subspecies of *c*, and *b* consists of those elements of *c* which cannot belong to *a*, then *a* consists of those elements of *c* which cannot belong to *b* (*complete principle of reciprocity of absurdity* or *principle of reciprocity of complementarity* or *complete principle of truth by non-contradictority*).

Another pair of corollaries of the simple and of the complete principles of judgeability respectively follows:

(1) every mathematical assertion can be *tested*, i.e. can either be proved to be non-contradictory or to be absurd (*simple principle of testability*, which is non-contradictory);

(2) if *a*, *b*, *d* and *c* are species of mathematical entities, and *a*, *b* and *d* are subspecies of *c*, and *b* consists of those elements of *c* which cannot belong to *a*, and *d* of those elements of *c* which cannot belong to *b*, then *c* is identical with the union of *b* and *d* (*complete principle of testability*, which is contradictory).

The assertion mentioned above of the existence of a sequence 0123456789 in the decimal expansion of π so far neither satisfies the simple principle of judgeability nor the simple principle of testability.

Let us consider a 'subordinating' sequence of *n* predicates of absurdity: *absurdity of absurdity ... of absurdity*, or in a shortened form, abs abs ... abs. The classical point of view admits the principle of reciprocity of complementarity, and thus allows this sequence to be reduced, by repeated cancellation of pairs of successive predicates, either to truth or to absurdity. One might expect that from the intuitionistic point of view such cancellations are strictly excluded, so that unequal sequences of this kind would have to be treated as inequivalent. But, surprisingly, this is not the case; cancellations of the kind mentioned are admissable, provided that they leave the first predicate of the sequence untouched, as follows from the following intuitionistic

THEOREM *Absurdity of absurdity of absurdity is equivalent to absurdity.*[11]

PROOF *Firstly,* since implication of the assertion y by the assertion x implies implication of absurdity of x by absurdity of y, the implication of *absurdity of absurdity* by *truth* (which is an established fact) implies the implication of *absurdity of truth,* that is to say of *absurdity, by absurdity of absurdity of absurdity. Secondly,* since truth of an assertion implies absurdity of its absurdity, in particular truth of absurdity implies absurdity of absurdity of absurdity.

From the theorem thus proved it follows that in intuitionistic mathematics every subordinating sequence of $n>2$ predicates of absurdity can be reduced either to absurdity or to absurdity of absurdity.

Another corollary of the same theorem is the intuitionistic validity of the principle of reciprocity of complementarity for negative assertions, which proves that this principle has a larger domain of validity than the principle of the excluded third.

By a *node of order n* we understand a sequence of n natural numbers ($n \geqq 1$) called the *indices* of the node.

A node p' of order $n+m$ ($m \geqq 1$) will be called an *mth descendant* of the node p of order n, and p will be called the *mth predecessor* of p', if the sequence of indices of p is an initial segment of the sequence of indices of p'.

The union of the node p and the species of its descendants will be called a *pyramid*, of which p will be called the top.

If $m=1$, p' will also be called an *immediate descendant* of p and p the *immediate predecessor* of p'.

The immediate descendants of a node p of order n in their natural order (i.e. ordered according to their last index) constitute a species Q of nodes. This species will be called a *row of nodes* of order $n+1$ and the *ramifying row* of p, whilst p will be called the *dominant* of Q.

The species of the nodes of order 1, in their natural order, will be called the *row of nodes of order* 1.

A finite sequence of nodes consisting of a node p_1 of order 1, an immediate descendant p_2 of p_1, an immediate descendant p_3 of $p_2, \ldots$ up to an immediate descendant p_n of p_{n-1}, will be called a *stick of order n.*

An infinite (but not necessarily predeterminate) sequence of nodes consisting of a node p_1 of order 1, an immediate descendant p_2 of p_1, an immediate descendant p_3 of p_2, etc., *ad infinitum*, will be called an *arrow*.

The arrow may proceed throughout with complete freedom, i.e. in the passage from p_ν to $p_{\nu+1}$ the choice of a new index to be joined to those of p_ν may be completely free for each ν as long as the creating subject likes. But this freedom of proceeding may at any stage be completely abolished at the beginning, or at any p_ν, by means of a law fixing all further nodes in advance. From this moment the arrow concerned will be called a *sharp arrow*. Finally the freedom of proceeding, without being completely abolished, may at some p_ν undergo some restriction, and later on further restrictions.†

All these intervening acts, as well as the choices of the p_ν themselves, may (at the beginning or at any later stage) be made

† In some former publications of the author restrictions of freedom of future restrictions of freedom, restrictions of freedom of future restrictions of freedom of future restrictions of freedom, and so on, were also admitted. But at present the author is inclined to think this admission superfluous and perhaps leading to needless complications. [A stronger statement can be found in Brouwer (1952).]

to depend on the influence of possible future occurrences in the world of mathematical thought of the creating subject.

Let ρ be a fundamental sequence $a_1,a_2,\ldots$ into which the species of nodes has been arranged in such a way that each node comes before its descendants, and before the nodes it precedes in its row of nodes. Then even if no details of this arrangement are known, the sequence of indices of each a_ν can be reconstructed as soon as for each a_ν its ramifying row can be indicated [including the row of nodes of order 1]. The arrangement can be effected, for example, in the following way.

Let G_n be the species of the nodes of order $\leqq n$ and indices $\leqq n$, $G_{n\nu}$ the species of the nodes of G_n of order ν, and A_n $(n \geqq 2)$ the species of the nodes of G_n not belonging to G_{n-1}. Each $G_{n\nu}$ is enumerated in such a way that p precedes q if the first index in which they differ is smaller for p than for q. Then each G_n is enumerated by making each $G_{n\nu}$ precede $G_{n,\nu+1}$. This implies an enumeration for each A_n. Finally the species of the nodes is enumerated by making G_1 precede each A_ν, and each A_ν precede $A_{\nu+1}$.

Let us suppose that in ρ an assignment is made to each node successively of either a 'figure', i.e. no thing or a mathematical entity previously acquired, or the predicate of being 'sterilized', in such a way that each descendant of a sterilized node is sterilized likewise, that the figures assigned in the case of non-sterilization are predeterminate (but not the decisions between sterilization and non-sterilization) and that for each non-sterilized node a non-sterilized immediate descendant can be indicated. Such a sequence of assignments will be called a *spread direction*.

If, instead of requiring that for each non-sterilized node a non-sterilized immediate descendant can be indicated, we impose the condition that for each n at least a non-sterilized node of order n is available, the sequence of assignments will be called a *spread haze direction*.

If the decisions of sterilization or non-sterilization are predeterminate, a spread direction will be called a *spread law*, and a spread haze direction a *spread haze law*.

A spread direction for which a non-sterilized node of order 1 can be indicated will be called *substantial*. If, however, all nodes of order 1 are sterilized, the spread direction will be called *empty*.

The part $\psi(\sigma,\chi)$ of a spread direction σ which concerns the pyramid χ will be called a *pyramid direction*, and a *pyramidal subdirection of* σ.

Each 'free arrow' of a spread direction σ, i.e. each arrow which avoids the nodes sterilized by σ, yields an infinite sequence of figures. These infinite sequences of figures, by virtue of their genesis, together with all infinite sequences equal to any one of them, may be considered as the elements of a species $\eta(\sigma)$. This species is called a *spread*.† The species of the infinite sequences derived in the same way from a spread haze direction is called a *spread haze*.

The existence of an element is guaranteed neither for a spread nor for a spread haze.

According to σ each pyramid χ yields a subspread $\eta(\sigma,\chi)$ of $\eta(\sigma)$. It is easily proved that the union of a finite number or a fundamental sequence of spreads is again a spread.

[In our discussion of the material of the Cambridge lectures we will use conventions and notations as presented in, for example, Kleene & Vesley (1965) or Troelstra (1977). These conventions deviate inessentially from Brouwer's in that zero is counted as a natural number, and that the empty sequence, $\langle\ \rangle$, is taken into consideration. The reader should keep these points

† Viewing the creation of the elements of the corresponding spread we can say that a *spread law* yields an *instruction* according to which, if again and again an arbitrary natural number is chosen as 'index', each of these choices has as its predeterminate effect, depending also on the earlier choices, that either a certain 'figure' (viz. either no thing or a mathematical entity previously acquired) is generated, or that the choice is 'sterilized', in which case the figures already generated are destroyed and generation of any further figures is prevented, so that all further choices are sterilized likewise. It was from this definition that intuitionist analysis started originally.

in mind when translating Brouwer's notions into the standard notation of contemporary practice.

We list the relevant notions:

(i) $x,y,z,\ldots$ vary over natural numbers;

(ii) $f,g,h,\ldots$ vary over lawlike functions from N to N (sharp arrows);

(iii) $\xi,\eta,\zeta,\ldots$ vary over choice sequences (arrows);

(iv) $n,m,\ldots$ vary over (codes of) finite sequences of natural numbers (nodes);

(v) the empty sequence $\langle\ \rangle$ (or λ) has code 0;

(vi) finite sequences will be written as $\langle n_0,\ldots,n_{k-1}\rangle$;

(vii) the concatenation function $*$ satisfies $\langle n_0,\ldots,n_{k-1}\rangle * \langle n_k,\ldots,n_p\rangle = \langle n_0,\ldots,n_p\rangle$ and $n*\langle\ \rangle = n = \langle\ \rangle * n$;

(viii) initial segments of functions or sequences are given by $\bar{f}x = \langle f0,\ldots,f(x-1)\rangle$, $\bar{\xi}x = \langle \xi 0,\ldots,\xi(x-1)\rangle$ and $\bar{f}0 = \langle\ \rangle = \bar{\xi}0$;

(ix) the length of a sequence is given by $l(\langle n_0,\ldots,n_{k-1}\rangle) = k$ and $l(\langle\ \rangle) = 0$;

(x) $n \preceq m$ if $\exists n'(n*n' = m)$, $n \prec m$ if $n \preceq m \wedge n \neq m$, $n \prec_1 m$ if $\exists x(n*\langle x\rangle = m)$, where $\preceq$ ($\prec$) is the predecessor, or successor, relation (strict predecessor, or strict successor, relation) and $\prec_1$ is the immediate predecessor, or successor, relation.

The species of all (codes of) finite sequences of natural numbers is denoted by SEQ.

In the Cambridge lectures Brouwer considered spreads in a more general setting. The natural denumeration ρ of all nodes is assumed, and by following it step by step one assigns figures to nodes, or sterilizes nodes. The sterilization is subject to the condition that the descendant of a sterilized node is itself sterilized. The figures assigned to nodes are predeterminate, i.e. no entities depending on choice sequences are allowed (higher order spreads were considered in Brouwer (1942b) and shown to be superfluous).

One can view the effect of sterilization as the specifying of a subtree (where trees are always considered to be closed under predecessors). Since the sterilization is, in general, not a predeterminate process, the resulting subtree of the universal tree may depend on choice parameters.

In a spread direction we have the strong condition $\forall n(\neg\mathrm{St}(n) \to \exists m \succ_1 n(\neg\mathrm{St}(n)))$, where St($n$) stands for '$n$ is sterilized'. We use St(n) instead of the more precise $\mathrm{St}_\sigma(n)$.

In a *spread haze* direction, however, we have the much weaker condition $\forall x \exists n(l(n)=x \wedge \neg\mathrm{St}(n))$. A spread haze direction thus can be, for example, well-founded, which is impossible for spread directions with at least one non-sterilized node.

In the definition of a spread direction nothing is said about the nodes of order 1. They may all be sterilized; then the spread direction will not contain an arrow (infinite path) so it is called empty. If there is a non-sterilized node of order 1, the spread direction is called *substantial.* Note that it is not possible to decide if a spread direction is empty or substantial. A substantial spread direction can be characterized by $\neg\mathrm{St}(\langle\ \rangle)$.

A *spread law* is a spread direction in which the decision between sterilization and non-sterilization is predeterminate, i.e. $\exists f \forall n(\neg\mathrm{St}(n) \leftrightarrow f(n)=0)$, so the subtree determined by the spread law has a lawlike characteristic function. This is the original notion of a spread (German: *Menge*), cf. Brouwer (1918), Kleene & Vesley (1965 p. 58), and Troelstra (1977 p. 127).

The notion of a spread was illustrated by Brouwer by the following example, taken from course notes of 1933. Consider two persons *A* and *B*. *A* calls out a natural number, which he has freely chosen. *B* has a stock of figures (signs), and according to a certain law *B* acts upon the number called by *A* in one of the following three ways.

(i) *B* writes down a figure and tells *A* to go on. *A* then calls out another natural number.

(ii) *B* writes down a figure and tells *A* that he may stop. Now *A* can stop or continue, but if *A* continues *B* does not write down any more signs.

(iii) *B* decides to destroy the results. *A* now stops and *B* destroys all the figures he has written down. We express this by saying that the choice sequence of *A* leads to sterilization (Dutch: *Stuiting*, German: *Hemmung*).

This is virtually the definition given in Brouwer (1918), where the notion was introduced. Note that the assignment of figures to sequences of numbers is supposed to be given by a law. The case presented under (ii) allows for the introduction of finite sequences. In the Cambridge lectures Brouwer replaced the halting by allowing B to write nothing (the empty figure). In (iii) the crucial notion of sterilization is introduced. One can think of this act of sterilization as a means of indicating subtrees of the universal tree of all finite sequences of natural numbers.

In 'Points and spaces' (Brouwer, 1954) Brouwer simplifies his presentation by dropping the sterilization clause. He achieves this either by allowing for each node all immediate descendants or by assigning a number m such that only immediate descendants with last index $\geqq m$ are allowed. Also in 'Intuitionism: an introduction' (Heyting, 1956) a notion of spread without sterilization is presented. This approach has been followed in all of the subsequent literature, e.g. Kleene & Vesley (1965), Troelstra (1969).

One might wonder why Brouwer chose to use the unwieldy notion of sterilization, whereas the 'subtree method' would do just as well. The reason is that Brouwer took the notion of mental, sequential process very seriously. For example, a construction is a mental process consisting of consecutive steps that might very well lead to an impossibility (for a geometric example see van Dalen, 1978), in which case the whole sequence is destroyed. Therefore the possibility of sterilization was built into his definition of spread. The intuitive picture is that of the chronicle of all construction attempts of a special kind (e.g. the construction of Cauchy sequences).

We proceed by exhibiting one of Brouwer's examples from his 1933 course.

> Construction of a spread consisting of three elements, such that the corresponding sequences of figures will be three given figures $\square$, x, σ.
>
> Consider the following law. If A presents the first time 1, 2 or 3 then B writes down $\square$, x or σ with the comment 'you may stop'. All other first choices of A lead to sterilization, i.e. destruction.

It is customary and convenient to represent spreads geometrically by pictures of trees. Strictly speaking there should always be two trees: one for the choice sequences of natural numbers, one for the associated sequences of figures. One often only considers the latter tree if it contains all the relevant information. We have added an extra node at the top of the tree in this example (Brouwer does not do this).

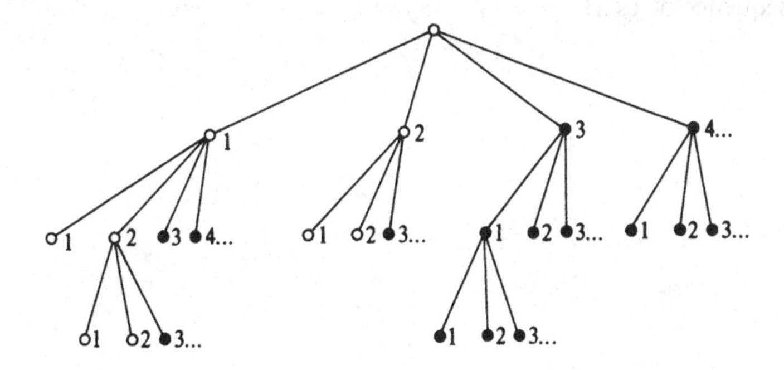

EXAMPLE Each node $\langle n_1,\ldots,n_k\rangle$ with $\max\{n_i|i\leqq k\}>2$ is sterilized. The resulting subtree of non-sterilized nodes is the binary tree. The sterilized nodes are indicated in the figure as black ones. Note that all descendants of black nodes are black. We now introduce an assignment:

$$\langle n_1,\ldots,n_k\rangle\mapsto\begin{cases}1 \text{ if } n_k \text{ is not preceded by any 2,}\\ \lambda \text{ else,}\end{cases}$$

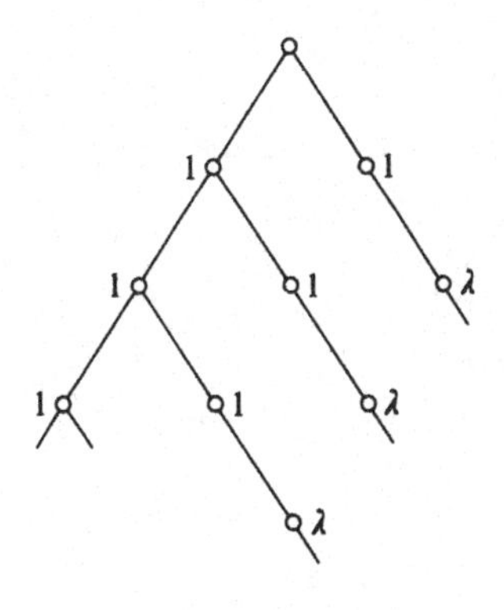

where λ is the empty word (or, in Brouwer's terminology, 'no thing').

We cannot make a tree of the generated sequences. The associated sequences can conveniently be represented by a labelled tree (p. 19). The resulting set of sequences consists of all sequences which have only initial segments of the form $\langle 1,1,1,\ldots,1\rangle$. (N.B. We cannot say 'all finite sequences $\langle 1,1,\ldots,1\rangle$ plus one infinite sequence of 1s'.)]

2

General properties of species, spread directions, spreads and spaces

If the species of the figures of a spread direction σ is discrete, σ and the spread deriving from it will be called *discernible*.

Two spreads will be said to be *concordant* if neither of them can possess an element distinct from all the elements of the other. Obviously two congruent spreads are also concordant, but the converse does not hold.

A spread direction σ will be called *blank* if to each non-sterilized node $a_1,a_2,\ldots, a_n$ it simply assigns its last index a_n.

A spread law σ will be called *full* if no sterilization occurs in it. The full blank spread law will be indicated by *Fb*.

A sterilized node of a spread direction σ will be called a *stop* if its immediate predecessor is non-sterilized.

Let K be the species of the nodes. Then a subspecies of K will be called *thin* if none of its nodes is a descendant of any other of its nodes. Thus the species $S(\sigma)$ of the stops of a spread direction σ, which is a removable subspecies of K, is also a thin subspecies of K.

Let $K(\sigma)$ be the species of the nodes which, according to the spread direction σ, are non-sterilized, or stops. A (not necessarily predeterminate) non-sterilized subspecies [i.e. consisting of non-sterilized nodes] $\beta(\sigma)$ of $K(\sigma)$ will be called a *barrage* of $K(\sigma)$ if no free arrow of σ can avoid it, i.e. if the species of the free arrows of σ passing through $\beta(\sigma)$ is congruent to the species of the free arrows of σ.[12] It will be called a *crude block* of $K(\sigma)$ if every free arrow of σ passes through it, i.e. if the species of the free arrows of σ passing through $\beta(\sigma)$ is identical to the species of the free arrows of σ. For instance such a crude block of σ will have to appear if we want to assign a natural number to *each* free arrow of σ, as the natural number in question will for each arrow have to be known at one of its nodes. These nodes form a species of non-sterilized nodes of σ through which each free arrow must pass.

Let $\beta^\circ(\sigma)$ be the union of a crude block $\beta(\sigma)$ of σ and the species $S(\sigma)$ of the stops of σ. Then $\beta^\circ(\sigma)$ will be a crude block of K. It will be called a *crude block completion* of $K(\sigma)$ and *the completion* of $\beta(\sigma)$.

The union of a crude block and the descendants of its elements, which may be regarded as a union of (not necessarily disjoint) pyramids, will be called a *massive crude block.*

A thin and removable (but not necessarily predeterminate) subspecies $B(\sigma)$ of $K(\sigma)$ will be called a *proper block*, or a *block* of σ if every free arrow of σ passes through it.

For such a block $B(\sigma)$ let $B'(\sigma)$ be the species of those stops of σ which are neither elements nor descendants of $B(\sigma)$. Then the union $B^\circ(\sigma)$ of $B(\sigma)$ and $B'(\sigma)$ will be a block of K. It will be called a *block completion* of $K(\sigma)$, and *the block completion of* $B(\sigma)$.

The nodes of $K(\sigma)$ which are not descendants of $B(\sigma)$ constitute a removable subspecies $B''(\sigma)$ of K. It will be called a *stump* of $K(\sigma)$, and we shall say that this stump is *carried* by the block $B(\sigma)$. This $B''(\sigma)$ is at the same time a stump of K carried by $B^\circ(\sigma)$. In $B''(\sigma)$ a predecessor of an element of $B(\sigma)$ may at the same time be a predecessor of an element of $B'(\sigma)$.

A node κ belonging to the stump τ together with its descendants belonging to τ constitutes a removable subspecies of τ which will be called a *stump sector* of τ, and will be said to be 'dominated' by its 'top' κ.

Extending the notions of a crude block, a crude block completion, a block, a block completion, a stump, and a stump sector from spread directions to pyramid directions, we remark that, χ being the pyramid corresponding to κ, the stump sectors of $K(\sigma)$ dominated by κ are identical to (pyramidal) stumps of $K\{\psi(\sigma,\chi)\}$.

The stump sectors of τ which are dominated by the elements of a row of nodes will be said to constitute, in their natural order, a *row of stump sectors.*

Let λ_ν be the node of order 1 and of index ν. The spread direction σ will be said to be *of clear opening* if either for each natural number n a natural number $\nu > n$ can be indicated such that λ_ν is non-sterilized, or a natural a_0 can be indicated such that the 'tailsegment' consisting of the λ_ν ($\nu > a_0$) is sterilized.[13]

⟦If, moreover, for each ν for which [the node of order 1 and index ν] λ_ν is sterilized, $\lambda_{\nu+1}$ is also sterilized, σ will be said to be of *condensed opening.*⟧[14]

If a natural number a_0 is indicated such that the tail consisting of the λ_ν ($\nu > a_0$) is sterilized, σ will be said to be *of limited opening.*

Let κ be a non-sterilized node of order m, and let κ_ν be its immediate descendant possessing ν as its $(m+1)$th index. Then, for the spread direction σ, κ will be said to be *of clear continuation* if either for each natural number n there is a natural number $\nu > n$ such that κ_ν is non-sterilized, or a natural number a_κ can be indicated such that the tailsegment consisting of the κ_λ ($\nu > a_\kappa$) is sterilized.

⟦If, moreover, for each ν for which κ_ν is sterilized, $\kappa_{\nu+1}$ is also sterilized, κ will be said to be of *condensed continuation.*⟧

If a natural number a_κ is indicated such that the tailsegment consisting of the κ_ν ($\nu > a_\kappa$) is sterilized, κ will be said to be of *limited continuation.*

A spread direction of clear opening, such that each non-sterilized node κ is of clear continuation, and such that for each κ the decision between the alternative ways of continuation is predeterminate, whilst if the second case occurs the choice of a_κ is also predeterminate, will be called *clear-cut.*

⟦A spread direction of condensed opening such that each non-sterilized node κ is of condensed continuation, and for each κ the decision between the alternative ways of continuation and if the second case occurs the choice of a_κ, are predeterminate, will be called *solid.*⟧

A spread direction of limited opening for which each non-sterilized node κ is of limited continuation, while each a_κ is predeterminate, will be called *bounded.*

A spread direction which is both solid and bounded will be called *firm.*

In a substantial clear-cut spread law σ let us say that in a row of nodes of order m the μth non-sterilized node possesses the number μ as its mth *rank number.* Let us delete in σ all sterilized nodes, and in each non-sterilized node replace the νth index by the νth rank number, for each ν. The result will be called a *spread key.*[15] Obviously a full spread law is at the same time a spread key.

Just as with the nodes of a spread direction, those of a spread key σ can in various ways be arranged into a fundamental sequence $a_1,a_2,\ldots$ in which each node comes before its descendants, and also before each node in the same row with higher last index.

Even if no further details of this arrangement are known, the sequence of indices of each a_ν can be reconstructed as soon as the row of nodes of order 1, and for each a_ν its ramifying row, can be indicated.

A spread key deriving from a bounded spread law will be called a *fan key,* and the corresponding spread will be called a *fan.* [A spread key derived from a firm spread law will be called a *cluster*, and the corresponding spread a *cluster spread.*]

If in the species of nodes of a spread key we sterilize a (not necessarily predeterminate) removable subspecies containing all descendants of its elements, such that a non-sterilized node of order 1 is given and for each non-sterilized node a non-sterilized immediate descendant can be indicated, the result will be called a *spread clue.*

A spread clue derived in this way from a fan key will be called a *bunch clue,* and the corresponding spread will be called a bunch.

We define a *pyramid key,* a *pyramidal subkey,* a *pyramid clue* and a *pyramidal subclue* in the same way as a pyramid direction and a pyramidal subdirection.

The notions of a crude block, a block and their completions, of a stump, a pyramidal stump and their sectors are easily extended from spread directions to spread keys and spread clues.

If M and N are disjoint subspecies of the series P and $M \cup N$ is congruent to P we shall say that P is *composed* of M and N,[16] and that M and N are *conjugate subspecies* of P. Thus, for instance, the species of exponents of Fermat's equation which render it soluble and insoluble respectively are conjugate subspecies of the species A of the natural numbers.

For a given P and any subspecies M a subspecies N can be indicated such that M and N are conjugate subspecies of P. However, this N is in general not uniquely determined by P and M. Thus for instance if P is the species of the real numbers and M the species of the irrational numbers, then for N we may choose the species of those real numbers whose rationality is non-contradictory as well as the species of the rational numbers.

If H and K are disjoint subspecies of the species P and $H \cup K$ is identical with P, so that H and K are conjugate removable subspecies of P, we shall say that P *splits* into H and K.[17] Thus, for instance, the species of the prime numbers and of the composite numbers are conjugate removable subspecies of A.

For an arbitrary proper subspecies H of P one cannot, in general, indicate a K such that H and K are conjugate removable subspecies of P. There are even infinite species which possess no removable proper subspecies at all.

If V and W are conjugate subspecies of P, and if in addition V consists of those elements of P which cannot belong to W, and W of those elements of P which cannot belong to V, we shall say that P is *directly composed* of V and W, and that V and W are directly conjugate subspecies of P. Thus, for instance, the conjugate species D and E consisting of those elements of P for which a certain negative property φ is true and absurd respectively are directly conjugate subspecies of P.

If the species P^2 of the pairs of elements of P is directly composed of a species of pairs of equal elements and a species of pairs of different elements, i.e. if within P non-contradictority of equality is equivalent to equality,† we shall say that P is *semi-discrete.*[19] Thus for instance each discernible spread is semi-discrete.

Let S be a species of subspecies s of the species P of which any two are disjoint. If P is congruent to $\cup S$ we shall say that P is composed of the species s, and that S is a species of *federate subspecies* of P.

Again, if P is identical with $\cup S$, we shall say that P splits into the species s.

Finally if P is composed of the species s, and if in addition each s consists of those elements of P which cannot belong to any other s, we shall say that P is *directly composed* of the species s, and that S is a species of *directly federate subspecies* of P.

If between two species M and N a (not necessarily predeterminate) one-to-one correspondence can be created, i.e. if to each element of M one element of N can be made to correspond in such a way that to equal and only to equal elements of M equal elements of N correspond, while to each element of N an element of M to which it corresponds can be indicated, we shall say that M and N are *equi-potential* and possess the same *power* or the same *cardinal number*.

Each finite number n as it is generated from the basic intuition gives rise to a *finite cardinal number* n.

Species which are equi-potential with a fundamental sequence will be called *denumerably infinite.*

A species which contains a denumerably infinite subspecies will be called *infinite.* In particular it will be called *reducibly infinite* if the denumerably infinite subspecies in question is removable. For instance the continuum, which will be intro-

†That this is not necessarily the case is shown by taking for P the species of all species of real numbers, and considering the following pair of elements of P: *firstly* the species of rational numbers; *secondly* the species of numbers of the form $r+s$, where s is a definite number whose rationality is non-contradictory although it is not rational, and r any rational number.[18]

duced later in this chapter, is infinite, but cannot possibly be reducibly infinite.

A species which cannot possibly possess a subspecies of the finite cardinal number n will be called *n-bounded*, and each species which is *n-bounded* from some n will be called *numerically bounded*.

A species which cannot possibly be infinite will be called *kept-down*.

A species which is equi-potential to a subspecies of a fundamental sequence will be called *denumerable*. In particular it will be called *countable* if the subspecies in question is removable. Thus, for instance, the species of the natural numbers that possess a given fleeing property is countable. On the other hand the species of the exponents of Fermat's equation that render it soluble is denumerable but not, as yet, countable.

In the case of denumerable or countable species it may happen that one can neither recognize it as empty nor point out one of its elements.

A spread direction will be called *individualized* if only to equal free arrows it assigns equal infinite sequences of figures, and *strictly individualized* if only to equal sticks it assigns equal finite sequences of figures.

THEOREM *Every discernible spread is identical with a subspecies of a spread deriving from a strictly individualized spread law.*[20]

To prove this theorem we start from the natural enumeration of the natural numbers; next we enumerate sequences of two natural numbers by the 'product' of the two natural enumerations of the natural numbers (by means of the diagonal process); then the sequences of three natural numbers by the product of the above enumeration of two natural numbers and another natural enumeration of the natural numbers (by means of the diagonal process); and so on. In this way in a spread law M a new index is assigned successively to each first choice, each second choice, third choice, etc. Then after any initial sequence of n choices ($n \geqq 0$), those natural numbers which do not already occur as choosable $(n+1)$th *new indices* are added as additional choosable $(n+1)$th *new indices*, all getting the sterilized character. In this way a new spread law N is generated, based on the *new indices*, which is identical with M and 'monotonic', i.e. (i) after an arbitrary non-sterilized nth index α only indices $\geqq \alpha$ can occur as non-sterilized $(n+1)$th indices; (ii) for each n no two equal non-sterilized nth indices can occur.

We now suppose that M is discernible and proceed to bring about an infinite sequence of changes $a_1, a_2, \ldots$ successively in the spread law N, such that for any n the initial columns of figures, generated by initial sequences of $n-1$ indices, are not changed by a_n.

By a_1 we first change in N the effect of the indices of the first choice in such a way that (i) every sterilized index remains sterilized, and (ii) a non-sterilized index ρ' becomes sterilized if there exists a lower index amongst the first choice which generates the same (i.e. an equal) figure. Furthermore let $\rho \neq \rho'$ be the lowest index amongst the first choices which generates the same figure as ρ', let $c(\rho')$ be an arbitrary non-sterilized initial index sequence in N which starts with ρ' and let $c(\rho)$ be the initial index sequence which is obtained from $c(\rho')$ by replacing its first index by ρ, and which therefore was a sterilized sequence in N. Then by a_1 the effect which $c(\rho')$ had in N is transferred to $c(\rho)$. Thus by a_1 N is changed into a monotonic spread law N_1, equal to N, but in which only equal first indices generate equal first figures.

By a_2 we first change in N_1 the effect of the indices of the second choice after each non-sterilized first index ρ in such a way that every sterilized index remains sterilized, and that a non-sterilized index τ' becomes sterilized if there exists a lower index of the second choice, coming after ρ, which generates the same figure. Again, let τ be the lowest index of second choice, coming after ρ, which generates the same figure as τ'. Let $c(\rho,\tau')$

be an arbitrary non-sterilized initial index sequence in N_1 starting with ρ,τ' and let $c(\rho,\tau)$ be the initial index sequence which is obtained from $c(\rho,\tau')$ by replacing its second index by τ and which therefore was a sterilized sequence in N_1. Then by a_2 the effect which $c(\rho,\tau')$ had in N_1 is transferred to $c(\rho,\tau)$. For every non-sterilized first index ρ of N_1 let $N_1(\rho)$ denote the spread law obtained from N_1 by compulsory first choice ρ and the destruction of its effect, i.e. the first figure. Then a_2 can be said to operate on each $N_1(\rho)$ just as a_1 operated on N. Thus by a_2 N_1 is changed into a monotonic spread law N_2, generating the same spread as N, but in which only equal first indices generate equal first figures, and only equal initial sequences of two indices generate equal columns of two figures.

For every non-sterilized initial sequence of two indices ρ,τ of N_2 write $N_2(\rho,\tau)$ for the spread law obtained from N_2 by compulsory first two choices ρ and τ and destruction of their effects. Then a_3 operates on each $N_2(\rho,\tau)$ just as a_1 operates on N and a_2 on each $N_1(\rho)$. Thus by a_3 N_2 is changed into a monotonic spread law N_3, generating the same spread as N_2, N_1 and N, but in which only equal first indices generate equal first figures, only equal initial sequences of two indices generate equal initial columns of two figures, and only equal initial sequences of three indices generate equal initial columns of three figures.

In general, let ν be any natural number. For any non-sterilized initial sequence of ν indices $\rho_1, \rho_2, \ldots, \rho_\nu$ define $N_\nu(\rho_1, \ldots, \rho_\nu)$ as the spread law obtained from N_ν by compulsory first ν choices $\rho_1, \rho_2, \ldots, \rho_\nu$ and the destruction of their effects. Then $a_{\nu+1}$ is to have the same effect on each $N_\nu(\rho_1, \ldots, \rho_\nu)$ as a_1 had on N. Hence N_ν is changed by $a_{\nu+1}$ into a monotonic spread law $N_{\nu+1}$ generating the same spread as N_ν, $N_{\nu-1}, \ldots, N_1$, N, in which, however, for each $\lambda \leqq \nu+1$, only equal initial sequences of λ indices generate initial columns of λ figures.

We remark that a non-sterilized initial sequence of m choices of indices $\leqq m$ in N_m corresponds uniquely to a non-sterilized initial sequence of m choices of the same indices in N_{m-1}, which generates the same figures. Similarly such a sequence corresponds uniquely to a non-sterilized initial sequence of m choices of indices $\leqq m$ in each of $N_{m-2}, N_{m-3}, \ldots, N_1, N$, all of which generate the same figures. Hence if we wish to determine the figure generated by, or (as the case may be) the sterilized conditions of, an initial sequence of m choices of indices $\leqq m$ in N_m, we need only consider successively the initial sequences of m choices of indices m in N, and hence this investigation is a finite process and therefore one which can actually be carried out. If therefore we define the spread law P by the requirement that, for each ν, initial sequences of ν choices of indices shall have the same effect in P as N_ν, then this gives us a complete definition for P.

Evidently, the spread law P is strictly individualized and generates every infinite sequence generated by N,† and hence every infinite sequence generated by M.

As an example, choose for M the individualized spread law depicted by the following diagram:[21]

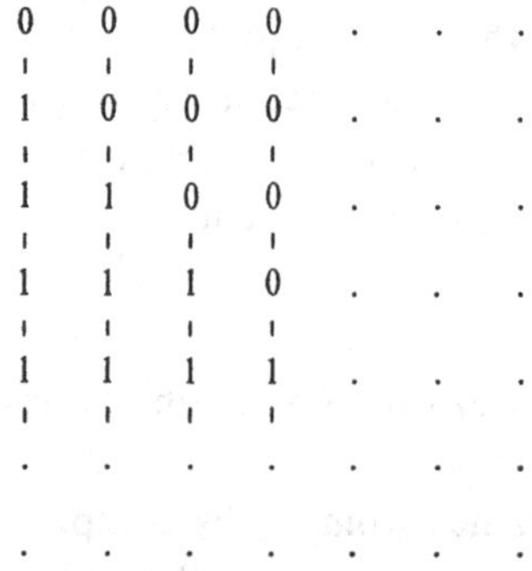

```
0   0   0   0   .   .   .
|   |   |   |
1   0   0   0   .   .   .
|   |   |   |
1   1   0   0   .   .   .
|   |   |   |
1   1   1   0   .   .   .
|   |   |   |
1   1   1   1   .   .   .
|   |   |   |
.   .   .   .   .   .   .

.   .   .   .   .   .   .
```

It is contained in the spread for P represented below, which is strictly individualized and has been derived from M in the way

† For, let $k_\nu(e)$ be the νth index of the infinite index sequences generating e in $N_\nu, N_{\nu+1}, \ldots$ Then the initial column of m figures of e is generated in N_m by $k_1(e), k_2(e), \ldots, k_m(e)$. Hence also in P the initial column of m figures of e is generated by $k_1(e), k_2(e), \ldots, k_m(e)$. And from this it follows that in P the infinite sequence $k_1(e), k_2(e), \ldots$ generates e.

described:

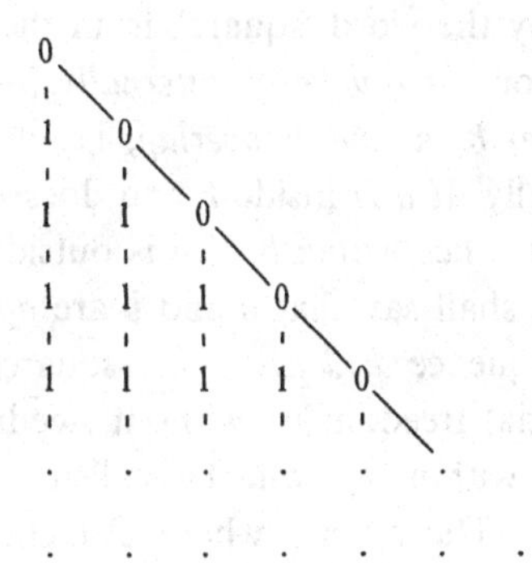

Evidently, P generates the infinite sequence of zeros which is not generated by M.

Now, however, let us suppose that M is bounded and assume that P generates an infinite sequence e which is distinct from any sequence generated by M. It follows from the main theorem of bounded spreads [the bunch theorem] (to be proved in Chapter 5) that a number n_e can be determined such that the initial sequence of n_e figures of e is distinct from all initial sequences generated by N_{v_e}, which is absurd. Since therefore our assumption is impossible, we have proved that a bounded discernible spread is contained in a concordant spread deriving from a strictly individualized spread law.

In the species of the binary fractions $a \cdot 2^{-m}$ arranged in their natural order (a and m integers), we consider, for given n, the pairs consisting of an element $a \cdot 2^{-(n+1)}$ and an element $(a+2) \cdot 2^{-(n+1)}$ and we call these pairs *$\lambda^{(n)}$-intervals*. By a *$\kappa^{(n)}$-interval* we understand a $\lambda^{(n)}$-interval for which a is even. All $\lambda^{(n)}$-intervals, for any n, will be called *λ-intervals*, and all $\kappa^{(n)}$-intervals, for any n, will be called *κ-intervals.*

By a *$\lambda^{(n)}$-square* we shall understand an ordered pair of $\lambda^{(n)}$-intervals. In particular, by a $\kappa^{(n)}$-square we shall understand an ordered pair of $\kappa^{(n)}$-intervals. The meaning of expressions for the relative situation of two λ-squares a and b derives in a natural way, suggested by the word 'square', from the natural order of the binary fractions: *a touches b internally* (*or externally*), *a is inside* (*or outside*) *b, a and b overlap*, i.e. cover one another entirely or partially. If a is inside b and does not touch b, then we shall say that a lies *within b*. If a is outside b and does not touch b, then we shall say that a and b are *apart*.

An infinite sequence $k_1, k_2, \ldots$ of λ-squares, proceeding in complete or partial freedom or without freedom, in which for each ν $k_{\nu+1}$ is within k_ν, will be called a *two-dimensional Cartesian point.*[22] The spread whose elements are these two-dimensional Cartesian points is called the two-dimensional Cartesian point spread. Denoting the two-dimensional Cartesian points $k'_1, k'_2, \ldots$ and $k''_1, k''_2, \ldots$ by p' and p'' respectively we shall say that p' and p'' coincide if it is certain that k'_μ and k''_ν overlap for each μ and ν. It is easily seen that coincidence is a transitive relation. We call the species of the two-dimensional Cartesian points which coincide with the two-dimensional point p a *two-dimensional Cartesian point core.* If, corresponding to a two-dimensional point core P, there exists a law $\sigma(P)$ which determines for each n a $\lambda^{(n)}$-square l_n, such that of each point $k_1, k_2, \ldots$ of P a computable tailsegment $k_m, k_{m+1}, \ldots$ is within l_n, then P is said to be a *sharp* two-dimensional Cartesian point core. The species of the two-dimensional Cartesian point cores is called the *two-dimensional Cartesian space*, or the *Cartesian plane*. The species of the sharp two-dimensional Cartesian point cores is called the *reduced two-dimensional Cartesian space* or the *reduced Cartesian plane.*[23] The points p' and p'' of the Cartesian plane† will be said to *deviate* if they cannot possibly coincide and to be *separated* if a square of p' and a square of p'' can be indicated which are outside one another. A point p *deviates* (*is separated*) from the species of point cores Q if it

† i.e. of point cores of the Cartesian plane. Similarly we shall speak of squares of point cores, meaning squares of points of point cores, and of squares of species of point cores, meaning squares of points of point cores of species of point cores.

deviates (is separated) from each point of Q. Thus for instance the 'origin' deviates from the species of irrational point cores and is separated from the species of those rational point cores which do not coincide with the origin.

Let p be a point and Q a species of point cores of the Cartesian plane. If within every square of p there lies a square of Q, p is said to be a *point of closure*, or a *limiting point*, of Q. The species of the points of closure of Q is called the *closure* of Q. If within each square of p there lie two mutually external squares of Q, p is said to be a *point of accumulation* of Q. The point core P belonging to the species of point cores Q is said to be an *isolated* point core of Q if a square k_n of P can be indicated such that no two mutually external squares of Q can lie within it. The point p will be called the *limiting point* of the (not necessarily discrete) infinite sequence $P_1, P_2, \ldots$ of point cores of the Cartesian plane if for each square λ_n of p a natural number m can be indicated such that for $\nu \geqq m$ P_ν lies within λ_n.

The point core P, belonging to the species of point cores Q, is said to be an *unapproachable* point core of Q if it cannot possibly be a point core of accumulation of Q.

The point core P belonging to the species of point cores Q is said to be a *covered* point core of Q if it cannot possibly be an isolated point core of Q.

Point cores of the Cartesian plane which cannot be point cores of accumulation of Q are called *unapproachable* by Q.

Two species of point cores of a Cartesian plane are said to be *locally concordant*, or shorter: *concordant*, if neither can contain a point core separated from the other. Thus, for instance, for an arbitrary species of point cores Q, the *closure* (i.e. the species of the point cores of closure) Q' is concordant with the union of Q and its *derivative* (i.e. the species of the point cores of accumulation) Q''. We show this by assuming that the point of closure p of Q is separated from all points of Q. Let k'_ν be any square of p within which the square k''_σ of the point q of Q lies. Since p and q

are separated, two squares k'_ρ $(\rho>\nu)$ and k''_τ $(\tau>\sigma)$ can be indicated which are mutually external and both within k'. But this holds for each ν, and hence we see that p is a point of accumulation of Q; thus Q' cannot contain a point which is separated from Q as well as from Q'', i.e. from $Q \cup Q''$.

Evidently concordance is a symmetric, reflexive and transitive property.

Two species of point cores of the Cartesian plane are said to be *locally congruent*, or shorter: *congruent*, if neither can contain a point core which deviates from the other. Thus, for instance, an arbitrary point core species Q is congruent with the union of the two species of its isolated and its covered points and also with the union of the species of its point cores of accumulation and its unapproachable point cores. Evidently congruence is a symmetric, reflexive and transitive property.

What has been said about the Cartesian plane can, *mutatis mutandis*, be carried over to n-dimensional Cartesian space $(n>2)$, where the analogues of squares will be called intervals of n-dimensions, and to one-dimensional Cartesian space, also called the *continuum*, whose point cores are called *real numbers.* And after having introduced real numbers we can assign n real numbers to each point core of the n-dimensional Cartesian space as their coordinates and a real number to each pair of point cores of that space as their Euclidean distance.

Returning to the Cartesian plane, let r be a binary rectangle defined on the analogy of the λ-squares, and let us understand by the *filling* of r the species of point cores of the Cartesian plane which cannot possess a square outside r. Then henceforth a binary square or a binary rectangle will mean the filling of a binary square or a binary rectangle in the sense hitherto used.

We shall call a point core species Q of the Cartesian plane *compact* if it is bounded (i.e. a λ-square L can be indicated outside which Q cannot possess a square) and closed (i.e. each convergent (not necessarily discrete) infinite sequence of elements of

Q possesses a limiting element belonging to Q). Furthermore a point core species Q of the Cartesian plane will be called *located* if its Euclidean distance to all binary point cores, and so to all point cores of the Cartesian plane, can be calculated with any degree of accuracy.

Let Q be a compact located point core species of the Cartesian plane, and L a λ-square outside which Q cannot possess a square.

In the following we shall understand by a $k^{(\nu)}$ a $\lambda^{(4\nu+1)}$-square ($\nu>0$) contained in L or overlapping L. Obviously we may suppose that for each ν the $k^{(\nu)}$ have been 'counted' as the elements of a finite species so that each $k^{(\nu)}$ has been given a natural number as its 'reference' number.

Each $k^{(\nu)}$ contains a ${}^0k^{(\nu)}$ and a ${}^1k^{(\nu)}$, both being squares concentric and homothetic with $k^{(\nu)}$ and having sides of length $\frac{3}{4}$ and $\frac{7}{8}$, respectively, of the side length of $k^{(\nu)}$. Then for each $k^{(\nu)}$ we can state either that it is *superfluous* for Q, i.e. that every point core of Q possesses a square lying apart from (i.e. external and not touching) ${}^0k^{(\nu)}$, or that it is *acceptable for* Q, i.e. that Q possesses a square lying within ${}^1k^{(\nu)}$.[24] We easily see that within each $k^{(\nu)}$ that has been stated acceptable for Q a $k^{(\nu+1)}$ can be indicated which *must* be stated acceptable for Q.

We now consider the nodes $a_1,\ldots,a_n$, where each a_ν is a reference number of a $k^{(\nu)}$, and the bunch σ acting on these nodes in the following manner:

(i) to each non-sterilized node $a_1,\ldots,a_\nu$ the $k^{(\nu)}$ having a_ν as its reference number is assigned;

(ii) the node a_1° is then and only then non-sterilized if the k' having this reference number has been stated acceptable for Q;

(iii) the node $a_1^\circ,\ldots,a_{\nu+1}^\circ$ ($\nu>0$) is then and only then non-sterilized, if (1) $a_1^\circ,\ldots,a_\nu^\circ$ is non-sterilized, (2) the $k^{(\nu+1)}$ having $a_{\nu+1}^\circ$ as its reference number has been stated acceptable for Q and lies within the $k^{(\nu)}$ having a_0° as its reference number.

The spread yielded by σ is a point spread s. Obviously every element of this bunch point spread s is a limiting point of Q and hence belongs to Q.

Conversely every point core P of Q contains a point of s. For, if for each ν we indicate a $k^{(\nu)}$, denoted by $k_P^{(\nu)}$, containing a square of P *within* its ${}^0k^{(\nu)}$ and therefore necessarily to be stated as acceptable for Q, then the infinite sequence $k'_P, k''_P, k'''_P, \ldots$ is assigned to a free arrow of σ (by σ), and consequently constitutes a point of s.

Thus we have proved:
Every compact located point core species of the Cartesian plane coincides with a bunch point spread.

Obviously the same theorem holds for compact located point core species of the n-dimensional Cartesian space and of the continuum.

In the special case that, in the Cartesian plane, Q is the *unity square* R, i.e. the $\lambda^{(0)}$-square with centre $(\frac{1}{2},\frac{1}{2})$, there is a test which for each $k^{(\nu)}$ decides unambiguously between the predicates *superfluous* and *acceptable*. For each $k^{(\nu)}$ either overlaps R, and then it is acceptable and not superfluous, or it lies outside R, and then it is superfluous and not acceptable. Hence in this case σ can be replaced by a *cluster* yielding the same spread s so that we have:

The unity-square of the Cartesian plane coincides with a cluster point spread.[25]

In the same way, μ being an arbitrary integer, we define a coincident cluster point spread for a *μ-system*, i.e. the closure of the union of a finite species of $k^{(\mu)}$-squares. For the points $p(k'_1, k''_2, \ldots)$ of this cluster point spread each $k^{(\nu)}$ is a $\lambda^{(4\nu+\mu+1)}$ square overlapping the μ-system. Here again the extension to the n-dimensional Cartesian space and to the continuum is obvious.

The infinite sequence of natural numbers passes into a located infinite species F,[26] if for any pair ν_1,ν_2 of its elements a *distance* $\rho(\nu_1,\nu_2)$ (i.e. a non-negative real symmetric function satisfying the usual triangle inequality) has been defined satisfying the following conditions.†

(i) For each n one can indicate a natural number μ_n such that $\rho(\nu,s_{\mu_n}) \leqq 4^{-n}$ for all $\nu \geqq \mu_n$, where $\rho(\nu,s_{\mu_n})$ is the minimum of $\rho(\nu,1),\ldots,\rho(\nu,\mu_n)$.[27] We shall express this by saying that F is approximated with any degree of accuracy by its successive initial segments $s_1,s_2,\ldots$

(ii) For each degree of approximation every element of the sequence is either *superfluous* (and then also superfluous for any higher approximation) or *acceptable*, i.e. for each n every element m of s_{μ_n} can be characterized either as a β_n-element or as an α_n-element. In the former case one can indicate a natural number $\mu(m,n)$ such that $\rho(\nu,m) \geqq \frac{5}{4}\cdot 4^{-n}$ for $\nu > \mu(m,n)$; in the latter case one can define an infinite sequence of different elements of F at a distance $\leqq \frac{3}{2}\cdot 4^{-n}$ from m. We shall say for each n that the β_n-elements are superfluous for the nth approximation and that the α_n-elements are acceptable for the nth approximation. The two alternatives are not mutually exclusive.

We shall say that an infinite sequence $a_1,a_2,\ldots$ of natural numbers is convergent with regard to the located infinite sequence F if for any n a natural number p_n can be indicated such that $\rho(a_\nu,a_\mu) < 2^{-n}$ for $\nu > p_n$ and $\mu > p_n$. Now we define a *point of* $R(F)$ as an infinite sequence of natural numbers containing an infinite number of different ones which converges with regard to F. If the convergent sequence concerned is predeterminate, we speak of a *sharp point* of $R(F)$.

Two points $p'(\nu'_1,\nu'_2,\ldots)$ and $p''(\nu''_1,\nu''_2,\ldots)$ of $R(F)$ are said to *coincide*, if for each n a natural number $\sigma(n)$ can be indicated such that any two elements of the species $\nu'_{\sigma(n)},\nu''_{\sigma(n)},\nu'_{\sigma(n)+1},\nu''_{\sigma(n)+1},$ $\nu'_{\sigma(n)+2},\nu''_{\sigma(n)+2},\ldots$ determine a distance $\leqq 2^{-n}$. The species of the

†For $\nu_1 \neq \nu_2$ a natural number $\psi(\nu_1,\nu_2)$ has to exist, such that $\rho(\nu_1,\nu_2)$ either $=0$ or $> 2^{-\psi(\nu_1,\nu_2)}$ and for $\nu_1 = \nu_2$ $\rho(\nu_1,\nu_2)$ has to vanish.

points of $R(F)$ which coincide with the point p of $R(F)$ is called a *point core* of $R(F)$. A point core of $R(F)$ is said to be sharp if it contains a sharp point. The species of the point cores of $R(F)$ is called the *located compact topological space* and is denoted by $R(F)$. The species of the sharp point cores of $R(F)$ is called a *reduced located compact topological space*, and will be denoted by $R^\circ(F)$.

$R(F)$ coincides with a bunch point spread whose elements are convergent infinite sequences of (not necessarily all different) elements of F, being successively an α_1-element, α_2-element, and so on, in such a way that the α_n-element α_n° may be followed by each α_{n+1}-element at a distance $\leqq 2\cdot 4^{-n}$ from α_n°, and by no α_{n+1}-element at a distance $\geqq 3\cdot 4^{-n}$ from α_n°.

The definitions, given above for n-dimensional Cartesian space, of deviation, separation, point of closure, point of accumulation, isolated point, unapproachable point, covered point, coincidence, concordance, congruence, closed, located and compact point species can easily be adapted to located compact topological spaces. For example, in $R(F)$ a point is a *point of accumulation* of the point core species Q if to each natural number n a natural number $\varphi(n)$ and two points p_1,p_2 of Q can be assigned such that $\rho(p,p_1)<2^{-n}$, $\rho(p,p_2)<2^{-n}$ and $\rho(p_1,p_2)>2^{-\varphi(n)}$.

The epithet 'located' for $R(F)$ itself is justified by the fact that the distance of $R(F)$ to any element of F can be calculated with any desired degree of accuracy.

The epithet 'compact' for $R(F)$ itself is justified by the fact that $R(F)$ is bounded and that for each convergent infinite sequence s of point cores $R(F)$ there exists a point core of closure belonging likewise to $R(F)$. Let $p',p'',\ldots$ be defining points of the elements of the sequence S. Let $p^{(n)}$ be the number sequence $a_1^{(n)},a_2^{(n)},\ldots$ Choose ρ_n so that for μ and $\nu \geqq \rho_n$, $\rho(p^{(\mu)},p^{(\nu)})<2^{-n-2}$. For each n choose a k_n so that $\rho(a_\mu^{(\rho_n)},a_\nu^{(\rho_n)})<2^{-n-2}$ for μ and $\nu \geqq k_n$. Now put $a_{\nu_n} \equiv a_{k_n}^{(\rho_n)}$, then for each $\mu, \nu \geqq n$ we have $\rho(a_{\nu_\mu}, a_{\nu_\nu}) \geqq 2^{-n}$. The point $p(a_{k_1}^{(\rho_1)}, a_{k_2}^{(\rho_2)}, \ldots)$ of $R(F)$ defines a limiting element of the original sequence S.

3

Order

Together with the empty twoity as the basic intuition of mathematics, 'between' was created, which is never exhausted by the interpolation of infinite sequences of new units. In Chapter 2 we have already recognized this intuitive 'between' as a matrix of 'point cores' and have joined it as such to intuitionist mathematics under the name of 'continuum'. In addition to this we shall now attempt to recognize the elements of order, naturally implied by the intuitive 'between', by seeking a definition for a natural order of the point cores of the continuum.

A species S will be said to be *partially ordered in projection*, if there are relations $x<y$, $x>y$, $x\mathrel{\underline{\underline{\circ}}}y$, each holding for certain 'ordered pairs' x,y of S and satisfying the following conditions:

(1) $a\mathrel{\underline{\underline{\circ}}}a$;
(2) $a\mathrel{\underline{\underline{\circ}}}b$ implies $b\mathrel{\underline{\underline{\circ}}}a$;
(3) $a\mathrel{\underline{\underline{\circ}}}b$ and $b\mathrel{\underline{\underline{\circ}}}c$ imply $a\mathrel{\underline{\underline{\circ}}}c$;
(4) $a<b$ and $b>a$ imply each other;
(5) $a<b$ and $a\mathrel{\underline{\underline{\circ}}}r$ and $b\mathrel{\underline{\underline{\circ}}}s$ imply $r<s$;
(6) $a<b$ and $b<c$ imply $a<c$;
(7) $a<b$ precludes $a>b$;

(In consequence of the above, $a\mathrel{\underline{\underline{\circ}}}b$ then precludes both $a<b$ and $a>b$, for, from $a\mathrel{\underline{\underline{\circ}}}b$ and $a<b$ would follow by (5): $a<a$, so by (4): $a>a$, which, by (7), is contradictory.)

The following statements will be equivalent: '$a<b$', 'a comes before b', 'b becomes after a', 'a is less than b', 'b is greater than a', 'a lies to the left of b', 'b lies to the right of a'. If for S equality of x and y (expressed by $x=y$) has been defined in such a way

that the relations $x = y$ and $x \doteq y$ are equivalent, S will be said to be *partially ordered*.

A partially ordered species S will be called (*simply*) *ordered* if for every pair of different elements a and b either $a < b$ or $b < a$, and *completely ordered* if it is ordered as well as discrete, so that for each pair of elements a and b either $a = b$ or $a < b$ or $a > b$. If between two (simply) ordered species a one-to-one correspondence obtains, we shall say that these species are *similar* and possess the same *ordinal number*.

Ordered species which are similar by means of predeterminate correspondence to the species of natural numbers in their natural order will be called *fundamental sequences*† and their ordinal number will be indicated by ω.

In the species of the binary fractions $a \cdot 2^{-m}$ in their natural order the ordinal number of the subspecies >0 and <1 will be indicated by η, and the ordinal number of the subspecies $\geqq 0$ and $\leqq 1$ by $\bar{\eta}$.

Now there corresponds to the sharp real number s_f, which was introduced in Chapter 1 (on a pre-intuitionist basis) by means of a fleeing property f, a point core of the reduced continuum introduced in Chapter 2, which demonstrates that the reduced continuum based on the intuitive 'between' is not completely ordered. Indeed, we can define a point $p(k_1, k_2, \ldots)$ of the reduced continuum coincident with s_f in the following manner: let, for any down-number n of f, k_n be a $\lambda^{(n-1)}$-interval symmetric with the origin core, while for any up-number n of f, k_n be a $\lambda^{(n)}$-interval symmetric with the point core $(-2)^{-\kappa_f}$. Then for this point p neither $p = 0$ nor $p < 0$ nor $p > 0$. The same holds for the point q, which is obtained in the way outlined for p, with the only difference that for an up-number n of f not $(-2)^{-\kappa_f}$ but $2^{-\kappa_f}$ is chosen as the centre of k_n. For this point q, moreover, $q < 0$ is absurd. If P is the subspecies of the continuum consisting of the rational numbers and the rational numbers to

† According to the classical meaning of the term, which justifies our using it here and there before.

which p has been added, Q the subspecies consisting of the rational numbers to which q has been added, then neither P nor Q is completely ordered; Q is ordered and P, as we shall see, is not.

More generally we can assert that, as long as a fleeing property exists such that neither the absurdity nor the absurdity of the absurdity of the existence of its critical number has been established, the reduced continuum is not discrete and therefore is not completely ordered.

But on the basis of the intuitive 'between' the reduced continuum is not ordered either. To see this, we consider a decomposition $\sigma(A)$ of the species of natural numbers A into two conjugate removable subspecies $\beta(A)$ and $\gamma(A)$, and a fleeing property f which is two-sided with respect to $\sigma(A)$, i.e. one for which neither the fact that κ_f cannot belong to $\beta(A)$ has been established, nor the fact that κ_f cannot belong to $\gamma(A)$ has been established, while also the absurdity of the absurdity of the existence of κ_f is supposed.[28] We define the point core $s(f,\sigma)$ as containing the following point $p'(k'_1,k'_2,\ldots)$ of the reduced continuum: let, for a down-number n of f, k'_n be a $\lambda^{(n-1)}$-interval symmetric with respect to the origin core, and let, for an up-number n of f, κ'_n be a $\lambda^{(n)}$-interval symmetric with respect to the point core $2^{-\kappa_f}$ if κ_f belongs to $\beta(A)$, and symmetric with respect to the point core $-2^{-\kappa_f}$ if k_f belongs to $\gamma(A)$.

Then $p' \neq 0$ (i.e. it is impossible that $p=0$), while neither $p'>0$ nor $p'<0$ is valid. Hence the reduced continuum is not ordered on the basis of the intuitive 'between' as long as there is no guarantee that for each fleeing property f, for which the absurdity of the absurdity of the existence of the critical number has been established, the following result is also valid: for each $\gamma(A)$ either the assertion 'κ_f belongs to $\beta(A)$' or the assertion 'κ_f belongs to $\gamma(A)$' can be proved to be absurd.

It will be clear from the above remarks on the reduced continuum that, *a fortiori*, ordering of the full continuum on the basis of the intuitive 'between' will be quite out of the question.

In particular, where in the reduced continuum of a point $p \neq 0$ such that neither $p > 0$ nor $p < 0$ is known we could only indicate the *absence of absurdity of existence* and the necessity of admittance of the possibility of existence (showing the *absence* of intuitive order in the continuum), in the full continuum we can prove the existence of such a point.

For this purpose we start from a mathematical assertion α such that no possibility of testing its validity (i.e. no method of deducing either its absurdity or the absurdity of its absurdity) is known, and define the following choices:

(i) as long as, during the process of choosing, neither the truth nor the absurdity of α has been established, we choose for k''_n a $\lambda^{(n-1)}$-interval which is symmetric with respect to the origin core;

(ii) if between the $(m-1)$th and the mth choice the truth of α has been established we choose for k''_n $(n \geqq m)$ a $\lambda^{(n)}$-interval symmetric with respect to the point core 2^{-m};

(iii) if between the $(r-1)$th and the rth choice the absurdity of α has been established, we choose for k''_n $(n \geqq r)$ a $\lambda^{(n)}$-interval symmetric with respect to the point core -2^{-r}.

The point core π_α containing the point $p''(k''_1, k''_2, \ldots)$ is indeed $\neq 0$ and neither <0, nor >0,[29] thus showing the absence of intuitive order in the full continuum as long as mathematical assertions are known for which no possibility of testing is known. [Added in pencil: the ordering, however, becomes contradictory if we consider the totality of all αs, each forming the rationality problem of a real number, and accordingly the totality of all p''s belonging to these αs.]

An important category of completely ordered species is formed by the *well-ordered species*.[30] Their introduction has to be preceded by the following definition.

Let a finite ordered species, or a fundamental sequence R, of disjoint completely ordered species N_ν be given, and let M be their union. M will be called the *ordinal sum* of the N_ν, if it has been completely ordered in such a way that, e' belonging to N' and e'' to N'', we have $e' < e''$ in M if either $N' < N''$ in R or $N' = N'' = N^\circ$ and $e' < e''$ in N°. The generation of this sum will be called *addition* and will be denoted in the usual way by the symbols $+$ and Σ.

Now a continually extending stock of well-ordered species is introduced according to the following rules.

(1) Each species containing one and only one element is a well-ordered species and, in this quality, will be called a *basic species.*

(2) If in the available stock of well-ordered species, a non-vanishing finite sequence of $n \geqq 1$ disjoint well-ordered species has been indicated, their addition will be called *first generating operation*, and their ordinal sum likewise becomes a well-ordered species and is added to the stock.

(3) If in the available stock of well-ordered species a fundamental sequence of disjoint well-ordered species has been indicated, their addition will be called *second generating operation*, and their ordinal sum likewise becomes a well-ordered species and is added to the stock.† [31]

A well-ordered species constructed without performing the second generating operation will be called *bounded.* Obviously a bounded well-ordered species is finite, which includes that its cardinal number can be calculated.

† As regards well-ordered species, ordinal numbers, at every moment only a finite species S_1 of symbols denoting them and only a finite species S_2 of symbols denoting laws of deduction of new ones from given ones, are available. But by means of S_1 and S_2 an infinite stock S_3 of well-ordered ordinal numbers and an infinite stock S_4 of laws of deduction of new well-ordered ordinal numbers from given ones can be constructed. And by devising appropriate laws of progression for fundamental sequences in S_3 and S_4 and performing the second generating operation, S_1 and S_2 can be extended.

Let F be a well-ordered species. All well-ordered species which have played a part in the above construction of F will be called *constructional* (or *hierarchical*) *subspecies* of F.

The constructional subspecies of F which have played a part in the *final* generating operation of F will be said to constitute the *row of constructional subspecies of order* 1 of F and will be denoted by F_ν (ν passing through an initial segment or through the whole of the species of natural numbers). The constructional subspecies of order 1 of F_{ν_1} will be said to constitute a *row of constructional subspecies of order* 2 of F and will be denoted by $F_{\nu_1\nu}$ (ν varying as above). And, in general, the row of constructional subspecies of order 1 of $F_{\nu_1\ldots\nu_k}$ will be said to constitute a *row of constructional subspecies of order* $k+1$ of F, and will be denoted by $F_{\nu_1\ldots\nu_k\nu}$ (ν varying as above). F itself will be considered as a *constructional subspecies of order* 0 *of* F.[32]

In this way each element, i.e. each basic species, of F and each constructional subspecies of F turn out to be constructional subspecies of finite order of F (which order, however, may indefinitely increase for appropriately chosen constructional subspecies). This property of F is easily proved by the *inductive method*, i.e. by remarking that it holds if F is a basic species, and that, when a generating operation is performed, it holds for the ordinal sum if it holds for its terms.

As a generalization of well-ordered species we introduce a continually extending stock of *marked well-ordered species* in the following way.

(1) To each element of a marked well-ordered species either the predicate 'full' or the predicate 'empty' has been assigned (the two predicates excluding each other). A marked well-ordered species in which a full element can be indicated will be called *substantial*, and one in which all elements are empty will be called empty.

(2) Each species containing one and only one element, with the predicate 'full' or 'empty', is a marked well-ordered species, and as such will be called a *basic species*.

(3) If, in the available stock of marked well-ordered species a non-vanishing finite sequence of disjoint marked well-ordered species is indicated, which is either substantial, i.e. in which a substantial element can be indicated, or is empty, i.e. all elements of which are empty, the addition of the elements of the sequence will be called *first generating operation*, and their sum likewise becomes a marked well-ordered set and is added to the stock.

(4) If in the available stock of marked well-ordered species a fundamental sequence of disjoint marked well-ordered species is indicated, which is either substantial, i.e. in which a substantial element can be indicated, or is empty, i.e. all elements of which are empty, the addition of the elements of the sequence will be called *second generating operation*, and their ordinal sum likewise becomes a marked well-ordered species and is added to the stock.

By means of the inductive method we can easily prove that each marked well-ordered species, and each of its constructional subspecies, is either substantial or empty.

The species of the full elements of a marked well-ordered species, conserving their original relations of order, will be called a *pseudo well-ordered species.* Pseudo well-ordered species are not, in general, well-ordered. For well-ordered species are, and pseudo well-ordered species are not, necessarily either finite or infinite. Simple examples of pseudo well-ordered species are furnished by substantial countable species.

If we consider the genesis of intuitionist mathematical deductions, we recognize that each mathematical deduction† M of a theorem can be mapped similarly [order-preserving] onto a pseudo well-ordered species $\mathscr{Q}(M)$ in such a way that all (full) basic elements of $\mathscr{Q}(M)$ are images of basic facts d, which are

†In general, philosophical considerations about mathematics do not have the character of mathematical deductions.

evident at once, and that the constructional subspecies of $\mathscr{P}(M)$ are images of deductions of intermediary theorems. If we denote by $a_{\nu_1 \ldots \nu_m}$ the intermediary theorem whose deduction is mapped onto the constructional subspecies $\mathscr{P}_{\nu_1 \ldots \nu_m}$ of $\mathscr{P}(M)$, then the generating operation creating $\mathscr{P}_{\nu_1 \ldots \nu_m}$ out of $\mathscr{P}_{\nu_1 \ldots \nu_m 1}$, $\mathscr{P}_{\nu_1 \ldots \nu_m 2}, \mathscr{P}_{\nu_1 \ldots \nu_m 3}, \ldots$ corresponds to the *final* inference stating $a_{\nu_1 \ldots \nu_m}$ as an immediate consequence of the union of $a_{\nu_1 \ldots \nu_m 1}$, $a_{\nu_1 \ldots \nu_m 2}, a_{\nu_1 \ldots \nu_m 3}, \ldots$ As regards the $a_{\nu_1 \ldots \nu_m}$ $(\nu = 1,2,\ldots)$, each of them may be included by any other of them, but they are all supposed to be deduced independently of each other.[33]

If the facts d and the inferences e, which thus correspond to (all full) basic elements of $\mathscr{P}(M)$ and to the generating operations of $\mathscr{P}(M)$ respectively, should at any time turn out to be decomposable into still more immediate facts d' and still more immediate inferences e', then for M a similar mapping on another pseudo well-ordered species $\mathscr{P}'(M)$ could be formed in such a way that the d' and e' would take over the part that the d and the e have played for the mapping $\mathscr{P}(M)$. [Written in pencil: rather deduce this analogous to Definitionsbereiche von Funktionen].

Further properties deduced by the inductive method are the following.

(1) The species $i_0(F)$ of the nodes belonging to the basic species of a well-ordered species F, and the species $i(F)$ of the nodes belonging to the constructional subspecies of F, are both removable subspecies of the node species K. Hence $i_0(F)$ and $i(F)$ are both countable. For the species of the nodes belonging to full basic species of a marked well-ordered species F^0 and the species of the nodes belonging to substantial constructional subspecies of F^0 the same property holds. Moreover, in the case of a well-ordered species F, $i_0(F)$ and $i(F)$ are both either finite or denumerably infinite.

(2) Let c be an element of the well-ordered species F other than

the first. Then the generation of F implies the generation of a well-ordered F_c, called a *stub* of F, and consisting of the elements of F preceding c in such a way that in F_c these elements have the same sequence of indices and the same natural relations of order as in F.

(3) A law which defines an element c' of the well-ordered species F, and assigns to each $c^{(\nu)}$ either the termination of the process or an element $c^{(\nu+1)}$ of F preceding $c^{(\nu)}$ in F, certainly indicates a natural number n and a corresponding element $c^{(n)}$ of F to which it assigns the termination of the process.

(4) For each well-ordered species F there is a one-to-one correspondence conserving the sequences of indices between the constructional subspecies of non-vanishing order of F on the one hand and the nodes of a spread key stump $\tau(F)$ on the other. In this correspondence the image of a generating operation of F in $\tau(F)$ is the passage of a row of nodes to its dominant, and the image of an element of F in $\tau(F)$ is an element of the block $\beta(F)$ carrying $\tau(F)$. A block and a stump of the character of $\beta(F)$ and $\tau(F)$ respectively will be called a *well-ordered block* and a *well-ordered stump* respectively. Both a well-ordered block and a well-ordered stump are removable subspecies of K.

Again, for each marked well-ordered species F there is a one-to-one correspondence, conserving the sequences of indices, between the constructional subspecies of non-vanishing order of F on the one hand, and the nodes of an (either substantial or empty) completed spread clue stump $\tau^{\circ}(F)$ on the other, this $\tau^{\circ}(F)$ deriving from a spread clue block $\beta(F)$ carrying a stump $\tau(F)$. In the said correspondence the image of a generating operation of F in $\tau^{\circ}(F)$ is the passage of a row of nodes to its dominant, while F itself is mapped onto the completed block $\beta^{\circ}(F)$ carrying $\tau^{\circ}(F)$ and splitting into the block $\beta(F)$ containing the images of the full elements of F, and a stop species $\beta'(F)$ containing the images of the empty elements of F. A completed spread clue block, a completed spread clue stump, a spread clue block, and a spread clue stump of the character of $\beta^{\circ}(F)$, $\tau^{\circ}(F)$, $\beta(F)$ and $\tau(F)$ respectively will be called a *well-ordered completed block*, a *well-ordered completed stump*, a *pseudo well-ordered block*, and a *pseudo well-ordered stump* respectively.

Obviously, if F is bounded, $\tau^{\circ}(F)$ will be a completed bunch stump.

The following classical theorems on well-ordered species are intuitionistically false.

(1) Any two well-ordered species are *comparable*, i.e. either they are similar or one of them is similar to a stub of the other.

Indeed, let f_1 and f_2 be two different fleeing properties such that between the existence of κ_{f_1} and the existence of κ_{f_2} no dependence is known. Let F_ν be a fundamental sequence for $\nu<\kappa_{f_1}$, but a single element for $\nu\geqq\kappa_{f_1}$, and G_ν a fundamental sequence for $\nu<\kappa_{f_2}$, but a single element for $\nu\geqq\kappa_{f_2}$. Then for the two well-ordered species $F=\Sigma_{\nu=1}^{\infty}F_\nu$ and $G=\Sigma_{\nu=1}^{\infty}G_\nu$ the above theorem does not hold.

(2) In a well-ordered species, each subspecies in which an element can be indicated contains a first element.

Indeed, let f be a fleeing property, and let $\kappa_f^{(\nu)}$ be the hypothetical νth natural number possessing f. We consider the well-ordered species $F+G$, where $F=\Sigma_{\nu=1}^{\infty}a_\nu$ and $G=\Sigma_{\nu=1}^{\infty}b_\nu$, each a_ν and each b_ν consisting of a single element. Then $\Sigma_{\nu=1}^{\infty}a_{\kappa_f}(\nu)+\Sigma_{\nu=1}^{\infty}b_\nu$ is a removable subspecies of $F+G$ for which the above theorem does not hold.

On investigating to what extent the continuum can be ordered on the basis of the intuitive 'between', we first remark that in any case we shall have to put $p'(k'_1,k'_2\ldots) < p''(k''_1,k''_2,\ldots)$ if some interval k'_n lies wholly to the left of some interval k''_ν. In this case we shall write $p \lessdot p''$, and say that p' lies sharply left of p''.

But the intuitive 'between' surely requires as well that the continuum contains further point cores between, for instance, the origin on the one hand and all positive rational point cores on the other. We find such points in the construction of the continuum described in Chapter 2, by considering the point spread for which only those intervals may be chosen which either have their centres in the origin or lie wholly to the right of the origin. This point spread coincides with the species of point cores which cannot be $\lessdot 0$, which we denote by saying that these point cores are $\geqq 0$. All those points of this spread which are prohibited from having the freedom of choice of their intervals entirely abolished at any stage or whose generation is subject to a prescription precluding the creation of a fundamental sequence of intervals with centres in the origin, do not just deviate from the origin, but are also distinguished from the origin by a quality of orientation to the right which necessitates their being labelled as 'to the right of the origin'. Hence the union of $\geqq 0$ and $\neq 0$ will have to be equivalent to > 0. As an example we consider the point $p^\circ(k^\circ_1,k^\circ_2,\ldots)$, determined by an assertion β, for which no possibility of testing is known. As long as this assertion has neither been proved correct nor absurd we choose for k°_n a $\lambda^{(n-1)}$-interval with centre at the origin. If between the choices of k°_{m-1} and k°_m either the correctness or the absurdity of β should have been established, then choose for k°_n $(n \geqq m)$ a $\lambda^{(n)}$-interval with centre at 2^{-m}. The point core ρ_β determined by this point p° is not $\gtrdot 0$, but it is $\neq 0$ and $\geqq 0$, hence > 0. Similarly, < 0 will have to be made equivalent to the combination of $\leqq 0$ and $\neq 0$. Evidently the combination of $\leqq 0$ and $\geqq 0$ is equivalent to $= 0$.

In an analogous manner it can be argued that if the continuum is being ordered on the basis of the intuitive 'between', the combination of $\geqq P$ and $\neq P$ will have to imply the property $> P$ for any point core P of the continuum. Hence $> P$ will have to be equivalent to the combination of $\geqq P$ and $\neq P$, and $< P$

with the combination of $\overset{\circ}{\leqq} P$ and $\neq P$, while again the combination of $\overset{\circ}{\leqq} P$ and $\overset{\circ}{\geqq} P$ is equivalent to $= P$.

If for two point cores A and B we denote the impossibility of $A > B$ by $A \leqq B$, then $A > B$ is implied by $A \overset{\circ}{>} B$ and therefore $A \overset{\circ}{\leqq} B$ by $A \leqq B$. But the converse also holds. For, since $A = B$ follows from $A \overset{\circ}{\leqq} B$ and $A \overset{\circ}{\geqq} B$, the simultaneous validity of $A \neq B$, $A \overset{\circ}{\leqq} B$ and $A \overset{\circ}{\geqq} B$ is absurd, whence it follows that $A \overset{\circ}{\leqq} B$ implies the absurdity of the combination of $A \neq B$ and $A \overset{\circ}{\geqq} B$, i.e. the absurdity of $A > B$ and hence the validity of $A \leqq B$.

We expect of the ordering relation $A < B$ introduced above, on the basis of the intuitive 'between' as the combination of $A \overset{\circ}{\leqq} B$ and $A \neq B$, that it should possess the property of transitivity, which is also required by the intuitive 'between'. This, indeed, is the case. For let us assume that $A < B$ and $B < C$. If now $A \overset{\circ}{>} C$ then natural numbers n, a and h could be indicated in such a way that some $\lambda^{(n)}$-interval $i^{(n)}(A)$ of A would lie to the right of a $\lambda^{(n)}$-interval $i^{(n)}(C)$ of C at a distance $\geqq a \cdot 2^{-h}$. An interval $i^{(n)}(B)$ of length $\leqq a \cdot 2^{-h-1}$ could, on account of $A < B$ and $B < C$, lie neither wholly to the left of $i^{(n)}(A)$, nor wholly to the left of $i^{(n)}(C)$, so that there would be no room for it. From this absurdity $A \overset{\circ}{\leqq} C$ follows. And if $A = C$, then we should reach the absurd conclusion that simultaneously $A = C > B$ and $A = C < B$. Hence $A \neq C$. And the combination of $A \overset{\circ}{\leqq} C$ and $A \neq C$ implies $A < C$. It can therefore be concluded that, on the basis of the intuitive 'between', there are relations $X = Y$, $X < Y$ and $X > Y$ holding for certain pairs X, Y of the continuum and possessing in addition to the properties formulated on p. 40 the following accessory properties.

(8) The simultaneous absurdity of $A = B$ and $A < B$ implies $A > B$.

(9) The simultaneous absurdity of $A < B$ and $A > B$ implies $A = B$.

A consequence of properties (8) and (9) is that no further assertions $X' = Y'$, $X' < Y'$ or $X' > Y'$ can be added in the continuum to the above system of relations $X=Y$, $X<Y$ and $X>Y$, introduced on the basis of the intuitive 'between', in such a way that the enlarged system is non-contradictory, i.e. that it is impossible to deduce contradictions from the enlarged system under application of the properties (1)–(7) inclusive.

For, suppose for instance that the relation $X'<Y'$ can be added in such a non-contradictory manner. Then the absurdity of $X'<Y'$ (on the basis of the intuitive 'between') would be absurd; consequently, by property (7), both $X'=Y'$ and $X'>Y'$ would be absurd and hence, by property (8), $X'<Y'$ would be valid, i.e. this relation was already present on the basis of the intuitive 'between'.

If in a species S criteria have been created for the relations $=$, $<$ and $>$, by virtue of which the properties (1)–(9) inclusive hold, we shall say that S is *virtually ordered*. In this case we have to understand the properties (2)–(9) in such a way that one or more possibilities or impossibilities of deducing one of these relations from the criteria imply another possibility or impossibility of deducing one of these relations from the criteria. [In an earlier version: ...in such a way that the possibility of deducing the validity (or absurdity) of one or more of these relations yields at the same time a possibility of deducing the validity (or the absurdity) of certain other relations.]

We can now state quite generally that the system concerned of relations in S is *inextensible*, i.e. that no further assertions $X'=Y'$, $X'<Y'$, or $X'>Y'$ (X' and Y' elements of S) can be added to the existing ones in a non-contradictory manner (which in this case means in such a way as to make it impossible to deduce contradictions from the old and the new relations under application of the properties (1)–(9).[34]

For, if for instance we have proved that the relation $X'<Y'$ can be added in a non-contradictory way, then by property (7) [and (6)] we have proved at the same time the impossibility of deducing $X'=Y'$ and $X'>Y'$ from the criteria and hence, by property (8), the possibility of deducing $X' < Y'$ from the criteria.

Conversely, if we have a species S in which, for the relations $=$, $<$ and $>$, criteria have been set up by virtue of which the properties (1)–(7) hold, and if the system of relations in S is inextensible, then the properties (8) and (9) also hold.

We restrict ourselves to the proof of property (9). We first remark that it will be sufficient to deduce from the simultaneous absurdity of the relations $X < Y$ and $X > Y$ (X and Y elements of S), by application of the properties (1)–(7), the possibility of adding the relation $X = Y$ to the existing ones in a non-contradictory manner (which in this case means in such a way as to make it impossible to deduce contradictions from the old relations and $X = Y$ under application of the properties (1)–(7)).

We therefore assume that $X > Y$ and $X < Y$ are absurd in consequence of the criteria of the given system of relations. Let α be the species of the old relations; G the species of elements π of S for which '$\pi = X$' or '$\pi = Y$' belongs to α; H the species of elements ρ of S for which '$\rho < X$' or '$\rho < Y$' belongs to α; K the species of elements σ of S for which '$\sigma > X$' or '$\sigma > Y$' belongs to α; β the species of assertions $\pi_\eta = \pi_\theta$, $\rho_\eta < \pi_\theta$, $\pi_\eta < \sigma_\theta$ and $\rho_\eta < \sigma_\theta$; γ the union of α and β.

We remark:

(i) G, H and K are disjoint. For, if there existed an element which belonged to two of these species then either $X > Y$ or $X < Y$ would follow.

(ii) In α no relations between a π and a π, between a ρ and a π, between a π and a σ, or between a ρ and a σ, other than those in the form of the assertions of β, are possible (for every relation of a different form between these pairs of elements would lead either to $X > Y$ or to $X < Y$). Hence the combination of a relation of α with an assertion of β can never yield a contradiction with property (7) [and (6)], and hence no pair of elements can occur in γ which yields a contradiction with property (7) [and (6)].

(iii) By applying properties (3), (4), (5) and (6) one can construct only relations of α out of relations of α; only assertions of β out of assertions of β; and only an assertion of β out of a relation of α

and an assertion of β (concerning the last part, the assertion $\rho_1 < \pi_1$ of β, for example, can only be combined with a relation of α in the following ways: with $\rho_\zeta = \rho_1$ to $\rho_\zeta < \pi_1$; with $\rho_\gamma < \rho_1$ to $\rho_\gamma < \pi_1$; with $\pi_1 = \pi_\eta$ to $\rho_1 < \pi_\eta$; and with $\pi_1 < \sigma_\theta$ to $\rho_1 < \sigma_\theta$). Hence applications of the properties (3), (4), (5) and (6) to elements of γ lead only to new elements of γ.

(iv) In particular, by combining α with the assertion $X = Y$ and applying properties (3), (4), (5) and (6), one can only deduce elements of γ, and therefore by (ii) one cannot deduce a contradiction in this way. Q.E.D.

Next we shall deal with the general problem of whether there is any possibility of ordering the continuum, or the reduced continuum, in a different way, be it on the basis of the intuitive 'between' or not.

For this purpose we shall assume for a moment that the (complete) continuum has been ordered in some way and we shall denote the corresponding relations of order by $\succ$ and $\prec$ respectively. Suppose that p_1 and v_1 are two different point cores of the continuum and that $p_1 \prec v_1$. Using the mode of measuring indicated by the binary fractions, we determine the point core w_1 'half way between' p_1 and v_1 by means of the λ-intervals of p_1 and v_1. Then we have either $w_1 \prec p_1 \prec v_1$ or $p_1 \prec v_1 \prec w_1$ or $p_1 \prec w_1 \prec v_1$. In the first case we put $w_1 = p_2$ and $v_1 = v_2$; in the last two cases we put $w_1 = v_2$ and $p_1 = p_2$. Hence in all cases we have $p_2 \prec v_2$, and the distance between p_2 and v_2 (using the mode of measuring indicated by the binary fractions) 'is half' that between p_1 and v_1. Proceeding in this way we construct for every natural number ρ the point core w_ρ 'half way between' p_ρ and v_ρ: we put $w_\rho = p_{\rho+1}$ and $v_\rho = v_{\rho+1}$ if $w_\rho \prec p_\rho \prec v_\rho$, and we put $p_\rho = p_{\rho+1}$ and $w_\rho = v_{\rho+1}$ if $p_\rho \prec v_\rho \prec w_\rho$ or $p_\rho \prec w_\rho \prec v_\rho$. In this way a fundamental sequence of pairs of point cores (p_ν, v_ν) $(\nu = 1,2,\ldots)$ is produced for which $p_\nu \prec v_\nu$ and the 'distance' between $p_{\nu+1}$ and $v_{\nu+1}$ 'is half' that between p_ν and v_ν.

With the aid of an assertion α for which no possibility of testing is known, we now define an infinitely proceeding sequence of pairs of point cores $(u'_1,u''_1),(u'_2,u''_2),\ldots$ in the following manner. As long as during the process of choosing successive pairs (u'_n,u''_n) neither the absurdity nor the absurdity of the absurdity of α has been established, we choose $u'_n=p_n$ and $u''_n=v_n$. If between the choice of (u'_{m-1},u''_{m-1}) and that of (u'_m,u''_m) the absurdity of the absurdity of α is established, then we choose all $u'_n=p_m$ $(n\geqq m)$ and all $u''_n=v_m$ $(n\geqq m)$. If between the choice of (u'_{r-1},u''_{r-1}) and that of (u'_r,u''_r) the absurdity of α is established, then we choose all $u'_n=v_r(n\geqq r)$ and $u''_n=p_r(n\geqq r)$. The sequence of pairs of point cores (u'_v,u''_v) then converges to a pair of point cores (u',u''). These point cores u' and u'' are *different*, for if they were equal absurdity and absurdity of absurdity of α would become absurd simultaneously, and this is patently absurd.

In consequence of our assumption, therefore, either $u'\prec u''$ or $u'\succ u''$ must hold. In the former case the absurdity of α, which was to imply $u'\succ u''$, proves to be absurd. In the latter case the absurdity of absurdity of α, which was to imply $u'\prec u''$, proves to be absurd, so that also α itself is absurd. In both cases a decision about α would have been arrived at, contrary to the assumption. Hence also each ordering of the (full) continuum which is unconnected with the intuitive 'between' involves a contradiction, as long as mathematical assertions are known, for which no method of testing is available [added in pencil: and if we take a variable α as the rationality problem of an arbitrary point of the continuum, then the impossibility appears that for *all* (u'_α,u''_α) a relation $\prec$ or $\succ$ can be fixed].

Now we put the same question with respect to the reduced continuum and we assume for a moment that it has been ordered in some way or other, and the signs $<$ and $>$ denote this order. Then any proof of the inequality of any two point cores of the reduced continuum would have to bring in its train an algorithm establishing between these two point cores one or the other of the relations $<$ and $>$.

To make the necessary deductions, consider the same indefinitely proceeding sequence of pairs of point cores p_ν, v_ν as above. Let $\sigma(A)$ be a decomposition of the species A of the natural numbers into two conjugate removable subspecies $\beta(A)$ and $\gamma(A)$, and f a fleeing property with critical number κ_f. We construct an indefinitely proceeding sequence of pairs of point cores (t'_ν, t''_ν) by putting $t'_n = p_n$ and $t''_n = v_n$ for a down-number n of f; by putting $t'_n = p_{\kappa_f}$ and $t''_n = v_{\kappa_f}$ for any up-number n of f, if κ_f belongs to $\beta(A)$; and by putting $t'_n = v_{\kappa_f}$ and $t''_n = p_{\kappa_f}$ for any up-number n of f, if κ_f belongs to $\gamma(A)$. This sequence of pairs of point cores (t'_ν, t''_ν) converges to a pair of point cores (t', t''). Now any proof of the absurdity of the absurdity of the existence of κ_f would have to imply the inequality of t' and t'' and hence, by the above, it would have to establish between t' and t'' one or other of the two relations $\prec$ and $\succ$. Hence it would have to entail a proof of the absurdity either of 'κ_f belongs to $\beta(A)$' or of 'κ_f belongs to $\gamma(A)$'.

But there exists no argument to show that every deduction of the absurdity of the absurdity of the existence of the critical number of a fleeing property establishes at the same time, for an arbitrary $\sigma(A)$, the absurdity of either the supposition that this critical number belongs to $\beta(A)$, or the supposition that it belongs to $\gamma(A)$. Thus also in the reduced continuum the existence of order, even independently of the intuitive 'between', is quite out of the question.

4

Precision analysis of the continuum†[35]

The introduction of the continuum on the basis of the λ-intervals of the binary fractions could create the impression that the numerical values of the binary fractions had some significance for this process. But in reality only a fundamental sequence φ was used as a basis, which in some way as an $\bar{\eta}(\varphi)$ had been completely ordered according to the ordinal number $\bar{\eta}$. In particular a '*closed continuum*' belonging to such an $\bar{\eta}(\varphi)$ is completely determined *independently* of a special similar mapping of $\bar{\eta}(\varphi)$ onto the species of the binary fractions $\geqq 0$ and $\leqq 1$ in their natural order (which similar mapping can be carried out in an infinity of ways). To see this we put $\{g_1,g_2,\ldots,g_\nu\}=\varphi_\nu$ [Brouwer writes '$\mathscr{P}(g_1,g_2,\ldots,g_\nu)=\varphi_\nu$'] and φ_ν, ordered after the everywhere-dense order of $\bar{\eta}(\varphi)$, will be indicated by $\bar{\eta}(\varphi_\nu)$. In $\bar{\eta}(\varphi)$ we perform an *intercalation subdivision*, i.e. by an indefinitely proceeding sequence (not necessarily predeterminate) of incorporations of the g_ν we construct a 'left' and a 'right' subspecies in $\bar{\eta}(\varphi)$; this means that every element of the left subspecies is to the left of every element of the right subspecies. The construction is carried out by determining the left and the right subspecies of $\bar{\eta}(\varphi_1),\bar{\eta}(\varphi_2),\bar{\eta}(\varphi_3),\ldots$ successively, where in each $\bar{\eta}(\varphi_\nu)$ *at most one* element g_{σ_ν} may remain outside the subdivision, and for each ν the element $g_{\sigma_{\nu+1}}$ (if it exists) is identical with g_{σ_ν} or with $g_{\nu+1}$. If $\psi(h_1,h_2,\ldots)$ is another enumeration of the elements of φ obtained by a one-to-one

† [Written in the margin, in pencil: The reduced continuum has measure 0. Measure of a point species = minimum measure if each point is replaced by one of its intervals. To modify along the lines of the Cape Town footnote. [cf. Brouwer (1952), p. 142. This modification concerns the first eleven lines of the chapter.]]

transformation of φ into itself, then an intercalation subdivision based on φ may as well be based on ψ. For, if we put $\{h_1, h_2, \ldots, h_\nu\} = \psi_\nu$, then for each m a natural number n can be indicated such that ψ_m is contained in φ_n.

We shall say that two intercalations of $\bar{\eta}(\varphi)$ *coincide* if never an element of the left subspecies of either of them can lie to the right of an element of the right subspecies of the other. Obviously any intercalation subdivision of $\bar{\eta}(\varphi)$ coincides with an intercalation subdivision of $\bar{\eta}(\varphi)$ in which g_{σ_ν} exists for every ν. The species of intercalation subdivisions t of $\bar{\eta}(\varphi)$ which coincide with an intercalation subdivision t_1 of $\bar{\eta}(\varphi)$ will be called an *intercalation place* of $\bar{\eta}(\varphi)$, and t and t_1 will be called *subdivisions belonging to this place.*

For the species E of the intercalation places e of $\bar{\eta}(\varphi)$ we construct a virtual order in the following manner. For the subdivision t' belonging to e' and t'' belonging to e'' we write $t' \lessdot t''$ if an element f' of the right subspecies of t' lies to the left of an element f'' of the left subspecies of t''. If then f_1 and f_2 (f_1 left of f_2) are two elements of $\bar{\eta}(\varphi)$ between f' and f'', and t'_0 is a subdivision which coincides with t', then f_1 (being to the right of f') cannot belong to the left subspecies of t', and hence f_2 belongs to the right subspecies of t'_0. Therefore it follows that $t'_0 \lessdot t''$, and since for any t''_0 which coincides with t'' we have likewise $t'_0 \lessdot t''_0$, we can write $e' \lessdot e''$. Furthermore we write $t' \underline{\lessdot} t''$ whenever $t' \gtrdot t''$ is absurd, in which case $t'_0 \gtrdot t''_0$ is also absurd, so that we may write $e' \underline{\lessdot} e''$. If t' and t'' coincide (and hence $e' = e''$) we write $t' \,\square\, t''$. If t' and t'' cannot coincide (and hence $e' \neq e''$) we write $t' \,\not\square\, t''$. If both $t' \underline{\lessdot} t''$ and $t' \,\not\square\, t''$ hold (and hence $e' \underline{\lessdot} e''$ and $e' \neq e''$), we write $t' < t''$ and $e' < e''$.

Since $t' \,\square\, t''$ is equivalent to the combination of $t' \underline{\lessdot} t''$ and $t' \underline{\gtrdot} t''$, the combination of $t' \,\not\square\, t''$, $t' \underline{\lessdot} t''$, $t' \underline{\gtrdot} t''$ is absurd. So $t' < t''$ precludes $t' > t''$, and conversely.

If we denote the absurdity of $t' > t''$ by $t' \leqq t''$, then $t' \leqq t''$ implies $t' \underline{\lessdot} t''$ since $t' \gtrdot t''$ implies $t' > t''$. But the converse also holds. For, since the combination of $t' \not\square\, t''$, $t' \underline{\lessdot}\, t''$, $t'' \geqq t'$ is absurd, it follows from $t' \underline{\lessdot}\, t''$ that the combination of $t' \not\square\, t''$ and $t' \geqq t''$ is absurd, i.e. that $t' \leqq t''$.

The transitivity of the relation $\lessdot$ is obvious.

To show that the relation $<$ likewise is transitive we assume $t' < t''$ and $t'' < t'''$ and suppose for a moment $t' \gtrdot t'''$. Then $t' \lessdot t''$ is impossible, because from $t' \lessdot t''$ and $t''' \lessdot t'$ would follow $t''' \lessdot t''$. So we would have $t' \geqq t''$, which would clash with $t' < t''$. Hence our supposition $t' \gtrdot t'''$ leads to a contradiction, so that we have deduced $t' \underline{\lessdot}\, t'''$ (1). Now let us suppose for a moment $t' \square\, t'''$. Then from $t' \square\, t'''$ and $t'' \underline{\lessdot}\, t'''$ would follow $t'' \underline{\lessdot}\, t'$, which would clash with $t'' > t'$. Hence our supposition $t' \square\, t'''$ leads to a contradiction, so that we have deduced $t' \not\square\, t'''$ (2). And from (1) and (2) it follows that $t' < t'''$.

Let a sequence of three successive elements of $\bar{\eta}(\varphi)$ be called a τ_n. A closed interval of $\bar{\eta}(\varphi)$ which for a certain τ_n is the smallest closed interval of $\bar{\eta}(\varphi)$ containing it, will be called a $\overline{\tau_n}$. An indefinitely proceeding (not necessarily predeterminate) sequence $g_{\alpha_1}, g_{\alpha_2}, \ldots$ of elements φ will be called *convergent* if to each natural number n a natural number m can be assigned such that for $\nu \geqq m$ all g_{α_ν} are lying within one and the same $\overline{\tau_n}$. Obviously we may then suppose that each $\overline{\tau_{n+1}}$ is a part of $\overline{\tau_n}$. Obviously the convergence property is invariant for one-to-one transformations of φ into itself. We shall call a convergent indefinitely proceeding sequence of elements of $\bar{\eta}(\varphi)$ a *limiting point* of $\bar{\eta}(\varphi)$. Two limiting elements $g_{\alpha_1}, g_{\alpha_2}, \ldots$ and $g_{\beta_1}, g_{\beta_2}, \ldots$ of $\bar{\eta}(\varphi)$ will be said to *coincide* if the infinite sequence $g_{\alpha_1}, g_{\beta_1}, g_{\alpha_2}, g_{\beta_2}, \ldots$ converges likewise. The definition just given passes into the one given on p. 33 by means of a similar mapping of $\eta(\varphi)$ onto the species of the rational numbers $\geqq 0$ and $\leqq 1$ in their natural order.

An intercalation subdivision t of $\bar{\eta}(\varphi)$ yields a limiting point $g_{\alpha_1}, g_{\alpha_2}, \ldots$ of $\bar{\eta}(\varphi)$ if for each g_{σ_ν} we choose the exceptional element g_{σ_ν}, if it exists, and otherwise the right end element of the left subspecies $\bar{\eta}(\varphi_\nu)$.

On the other hand a limiting point s' of $\bar{\eta}(\varphi)$, if the middle elements of its successive $\overline{\tau_\nu s}$ are chosen as g_{σ_ν}s, generates an intercalation subdivision t' of $\bar{\eta}(\varphi)$ (and coinciding limiting points s' generate coinciding subdivisions t'), while t' in the above way generates a limiting point s'' coinciding with s'.

We shall say that a point p of the continuum and an intercalation subdivision of $\bar{\eta}(\varphi)$ *coincide*, if it is impossible that an element of φ belonging to the right (left) subspecies of t is to the left (right) of an interval of p. We shall say that a point core P of the continuum and an intercalation place e of $\bar{\eta}(\varphi)$ coincide if a point of P coincides with a subdivision of e, and hence all points of P coincide with all divisions belonging to e. The correspondence between the continuum and the species of intercalation places of $\bar{\eta}(\varphi)$, determined by the relation of coincidence, leaves the relations expressed by the signs $=$, $\lessdot$ and $\gtrdot$ invariant, and thus also the relations expressed by the signs $<$ and $>$. Hence this correspondence is a similarity, and one sees that the continuum has been reconstructed independently of the species of binary fractions.

Let ξ be a denumerable infinite species of intercalation places of $\bar{\eta}(\varphi)$, which is itself of type η and everywhere-dense in $\bar{\eta}(\varphi)$ while any two of its elements are also *apart* (i.e. separated by at least two elements of φ); we shall say that ξ is a *deformation* of $\bar{\eta}(\varphi)$. Then $\eta(\varphi)$ is also a deformation of ξ. If we define the coincidence of an intercalation subdivision of ξ with an intercalation subdivision of $\bar{\eta}(\varphi)$ in the same way as the coincidence of two intercalation subdivisions of $\bar{\eta}(\varphi)$, a similarity between the 'continuum over $\bar{\eta}(\varphi)$' and the 'continuum over ξ' is produced.

Thus the species of the binary fractions, of the decimal fractions, of the rational numbers, and of the algebraic numbers appear as examples of creatively equivalent 'skeletons' of one and the same continuum.

If e is an element of the continuum, and f' and f'' are two *different* elements of the *same* skeleton s of the continuum, then we have either one of the relations $e \leqq f'$ and $e \geqq f'$, or one of the relations $e \leqq f''$ and $e \geqq f''$. Similarly, given the element e of the continuum and an arbitrary finite number of different elements $f', f'', \ldots, f^{(\nu)}$ of s, we have for $1 \leqq \nu \leqq m$, with at most one exceptional value of ν, one of the two relations $e \leqq f^{(\nu)}$ and $e \geqq f^{(\nu)}$. If e is such that for any m and any choice of the $f^{(\nu)}$ there exists no such exceptional value ν, then we shall say that e possesses in s a *first-order precision position.*[36] As an example of an element of the continuum which does not possess this precision position with regard to any of the above skeletons, we can cite the two-sided (with respect to parity) binary shrinking number s_f, introduced in Chapter 1. Precision position of the first order is the necessary and sufficient condition that e determines in s a partition which is a Dedekind cut ['incision' in the manuscript]. Thus we see that the Dedekind cut is inadequate as a mechanism for the definition of a general element of the continuum, and even of the reduced continuum.

If the species of decimal fractions is employed as a skeleton s, the precision position of the first order of the element e of the continuum can be interpreted as the necessary and sufficient condition for the expansibility in an infinite decimal fraction.[37] For, an infinite decimal fraction with which e coincides can be formed if and only if for *each* decimal fraction d either $d \geqq e$ or $d \leqq e$ is known to hold. Since, however, for a given d and a given e *both* relations $d \geqq e$ and $d \leqq e$ may be found to hold, the infinite decimal fraction belonging to e need not be uniquely determined.

If, for e and every element f of s, one of the relations $e < f$ and $e \geqq f$ holds we shall say that e possesses in s a *second-order precision position of the first kind*, and, if always either $e \leqq f$ or $e > f$, a *second-order precision position of the second kind.*

If the species of decimal fractions is employed as a skeleton s, the second-order precision position of the first kind of the element e of the continuum can be interpreted as a necessary and sufficient condition for the expansibility in a unique infinite decimal fraction, with the exclusion of the occurrence of a last digit, different from 9. Similarly, precision position of the second order and of the second kind is necessary and sufficient for expansibility in a unique infinite decimal fraction, with the exclusion of the occurrence of a last digit other than 0.

Next we choose for s the species of the rational numbers and investigate the conditions under which the element e of the continuum can be expanded in a regular continued fraction. If such an expansion is possible then for each approximating fraction f_n of e it must be known firstly that either $e=f_n$ or $e \neq f_n$, secondly, in the latter case, that either $e>f_n$ or $e<f_n$, and thirdly, again in this case, that either $e \gtrdot f_n$ or $e \lessdot f_n$ (because the calculation is only possible if we can indicate an upper bound for the natural number which is to occur next in the expansion). But since the approximating fractions of the expansion of an element e of the continuum in a regular continued fraction form a removable subspecies of the species of rational numbers, and since each rational number which is not an approximating fraction is separated from e by an approximating fraction, it then follows necessarily that for each rational number f either $e=f$ or $e \gtrdot f$ or $e \lessdot f$ is known to hold.

From the above one sees how much more special the elements of the (reduced as well as full) continuum are that are expansible in a regular continued fraction than those expansible in an infinite decimal fraction.

Since most of the classical properties of the continuum are based on the assumption that the continuum is completely ordered we cannot expect that they will not carry over in the old form. We shall start by enumerating these classical properties[38] and then we shall verify them.

(1) THE ORDINAL DENSITY-IN-ITSELF The element a of a completely ordered species S is an *ordinal limit of an increasing infinite sequence* $a_1, a_2, \ldots$ of elements of $S (a_1 < a_2 < a_3 < \ldots < a)$, if for every $b < a$ there exists an $a_n > b$. An *ordinal limit of a decreasing infinite sequence* is defined in an analogous manner. If every element of a completely ordered species S is an ordinal limit [in both senses], then S is said to be *ordinally dense-in-itself.*

(2) THE ORDINAL SEPARABILITY-IN-ITSELF A completely ordered species S is said to be *ordinally separable-in-itself,* if in S a completely ordered fundamental sequence F can be indicated such that between any two different elements of S there is an element of F.

(3) THE ORDINAL CONNECTEDNESS To define this notion for completely ordered species we shall, as before, call two subspecies α and β of the completely ordered species S *ordinally separate subspecies* if every element of α is to the left of every element of β. Then the completely ordered species S is called *ordinally connected* if in each ordinal separation of S (i.e. in each subdivision of S into two ordinally separate subspecies α and β) either α contains a rightmost and β no leftmost element, or α contains a leftmost and β no rightmost element.

(4) THE ORDINAL EVERYWHERE-DENSITY A completely ordered species S is said to be *ordinally everywhere-dense* if for each two different elements a and b of S $(a < b)$ there exists an element c of S such that $a < c < b$.

(5) THE ORDINAL LOCAL COMPACTNESS A completely ordered species S is said to be *ordinally locally compact* if for each infinite sequence $I_1, I_2, \ldots$ of closed intervals of S which is *telescopic* (i.e. such that for each ν $I_{\nu+1}$ is a subspecies of I_ν), there exists an element of S common to all I_ν.

[Investigation of the properties of the continuum:]

[THE ORDINAL DENSITY-IN-ITSELF] Let us suppose that in the intuitionist reduced continuum the real number 0 (zero) is an ordinal limit of the decreasing sequence $a_1,a_2,\ldots$ Let f be a fleeing property, and let us suppose that for the existence of κ_f the absurdity of the absurdity has been established. Let $d_\nu = a_\nu$ for each down-number ν of f and $d_\nu = a_{\kappa_f}$ for each up-number ν of f. Then the real number d, to which the infinite sequence $d_1,d_2,\ldots$ converges, is >0, but we cannot indicate an $a_\nu < d$. Thus the property of being an ordinal limit in the intuitionist reduced continuum cannot possibly have been established for the real number 0, and the reduced continuum cannot be ordinally dense-in-itself, as long as we cannot exclude fleeing properties for which the absurdity of the existence of a critical number is absurd.

Again let us suppose that in the intuitionist full continuum the real number 0 is an ordinal limit of the decreasing sequence $b_1,b_2,\ldots$ Let α be an assertion for which no method of testing (i.e. of deducing either its absurdity or the absurdity of its absurdity) is known. Let us choose $e_\nu = b_\nu$ as long as neither the absurdity nor the absurdity of the absurdity of α has been established, and $e_\nu = b_m$ for $\nu \geqq m$ if between the choice of e_{m-1} and that of e_m either the absurdity or the absurdity of the absurdity of α has been established. Then the real number e, to which the infinite sequence $e_1,e_2,\ldots$ converges, is >0, but we cannot indicate a $b_\nu < e$. Thus the assertion that the real number 0 is an ordinal limit in the intuitionist full continuum and the assertion that the intuitionist full continuum is ordinally dense-in-itself, are contradictory, as long as there exist mathematical assertions for which no method of testing is known.

But we can adapt the notion of ordinal density-in-itself to intuitionism, i.e. bring it into a form which is equivalent to the original one in the classical, but not in the intuitionist, sense, and such that it is satisfied by the intuitionist continuum.

To this end the notion of an interval is extended to virtually ordered species in the following way: if a and b are two arbitrary elements of the virtually ordered species S, the species of those elements c of S for which neither the combination of the relations $c>a$ and $c>b$ nor the combination of the relations $c<a$ and $c<b$ can be valid will be called the *closed interval ab*. Furthermore we shall say that the element g of S lies *between* the elements a and b of S if it is different from both a and b, and it belongs to the closed interval ab. In the case $a<b$ this 'between' is equivalent to the classical 'between'. The species of those elements which lie between a and b will be called the *open interval ab*. The elements a and b will be called the *end elements* of the open as well as of the closed interval ab. Now we shall call the element e of S a *condensation element* of S, if there exists an infinite sequence of closed intervals of S, each of which is a proper subspecies of the preceding one and contains e, while each element of S which belongs to all intervals of the sequence is equal to e. If we call a virtually ordered species *ordinally dense-in-itself in the wider sense*, if each of its elements is a condensation element, then this definition is equivalent to that of ordinal density-in-itself in the classical sense, while both the intuitionist full continuum and the intuitionist reduced continuum are ordinally dense in the wider sense.

For a closer consideration of the intervals of the full continuum we shall introduce a *quartering point* of the continuum, as a point $(k_1,k_2,\ldots)$ where k_ν is a $\lambda^{(4\nu+1)}$-interval. Let $\alpha(k'_1,k'_2,\ldots)$ be a quartering point of the point core a, $\beta(k''_1, k''_2,\ldots)$ a quartering point of the point core b, and let, for each ν, the $\lambda^{(4\nu+1)}$-interval k'''_ν be one of the intervals k'_ν and k''_ν and k^{IV}_ν the other, in such a way that the centre of k^{IV}_ν never lies to the left of the centre of k'''_ν. Then, for each ν, $k'''_{\nu+1}$ lies within k'''_ν and $k^{\mathrm{IV}}_{\nu+1}$ within k^{IV}_ν. We see that this property would need a proof only in the case that k'''_ν and k^{IV}_ν overlap without coinciding, while k'''_ν and $k'''_{\nu+1}$ (thus also k^{IV}_ν and $k^{\mathrm{IV}}_{\nu+1}$) belong to different elements of the pair (α,β), so that both $k'''_{\nu+1}$ and $k^{\mathrm{IV}}_{\nu+1}$ lie within the 'common part' of k'''_ν and k^{IV}_ν. But also in this case the property is obvious.

Let c be the point core defined by the point $\gamma(k_1''',k_2''',\ldots)$ and d the point core defined by the point $\delta(k_1^{\mathrm{IV}},k_2^{\mathrm{IV}},\ldots)$. Then $c \leqq d$, and c and d, which obviously are fully determined by a and b, will be the *left* and *right limiting points* of the open as well as of the closed interval ab. Let us suppose that c would be $\neq a$ as well as $\neq b$. Then, for every ν, k_ν' and k_ν'' must overlap, for, if for any ν they did not overlap, c would coincide either with a or with b. But if, for every ν, k_ν' and k_ν'' overlap, then $a=b$ and $c=a=b$, which contradicts our supposition. So we have established that *neither of the limiting points of the interval ab can possess the property of being simultaneously* $\neq a$ *and* $\neq b$.

Let σ_ν be the species of $\lambda^{(4\nu+1)}$-intervals with centres lying neither to the left of the centre of k_ν''' nor to the right of the centre of k_ν^{IV}, and let s_ν be the union of the closed intervals of the continuum corresponding to the elements of σ_ν. If a and b are sharp point cores, the infinite sequence $s_1,s_2,\ldots$ can be constructed as a predeterminate one. The closed interval cd is identical with the intersection of $s_1,s_2,\ldots$, which in its turn is easily seen to coincide with the cluster spread C whose elements are the quartering points $(k_1,k_2,\ldots)$, where each k_ν is an element of σ_ν. We shall show that also the closed interval ab coincides with C.

To this end we shall prove first of all that an arbitrary element e of C cannot possibly be simultaneously $>a$ and $>b$. Indeed, let us suppose that e would be $>a$ and $>b$. Then the relations $e \lessdot a$ and $e \lessdot b$ would both be impossible. Thus for an arbitrary k_ν of e it would be impossible to lie apart to the left both from k_ν' and k_ν'', and impossible as well to lie apart to the left from k_ν^{IV}, so that the relation $e \geqq d$ would hold. On the other hand we know that for every element e of C the relation $e \leqq d$ is valid. So our supposition leads to $e=d$, thus to the simultaneous validity of $d>a$ and $d>b$, which clashes with the above theorem that the combination of $d \neq a$ and $d \neq b$ is impossible. Thus we have proved the absurdity of the simultaneous validity of $e>a$ and $e>b$, and we can prove the absurdity of the simultaneous validity of $e<a$ and $e<b$ in the same way, so that C coincides with a subspecies of the closed interval ab.

On the other hand let $\pi(k_1,k_2,\ldots)$ be a quartering point of the closed interval *ab*. Then each $k_{\nu+1}$ must overlap $\sigma_{\nu+1}$, which means that it must be within σ_ν. So π belongs to s_ν for each ν, i.e. to the intersection of $s_1,s_2\ldots$ Hence it coincides with a point of *C*.

We have now established that every closed interval is identical with a *binary lump*, which is the intersection of a convergent telescopic infinite sequence of *binary* closed intervals. The converse also holds, because the sequence of end points, corresponding to the sequence of intervals defining a binary lump ρ, converges to the point cores *c* and *d* ($c \leqq d$) such that each point core of ρ is $\geqq c$ and $\leqq d$, while each point core which is simultaneously $\geqq c$ and $\leqq d$ belongs to ρ.

[THE ORDINAL SEPARABILITY-IN-ITSELF] Let us suppose that *F* is a completely ordered fundamental sequence demonstrating the separability-in-itself of the intuitionist reduced continuum. As this is certainly impossible if 'between' is understood in its classical sense, we shall take this concept here also in the sense explained in the investigation of ordinal density-in-itself. Let p_1 and p_2 be two arbitrary elements of *F* such that $p_1 < p_2$ according to the virtual order of the reduced continuum. Let p_3 be the first element of the fundamental sequence *F* which lies between p_1 and p_2, according to the virtual order of the reduced continuum, p_4 the first element of the sequence which lies between p_1 and p_3 according to the virtual order of the reduced continuum, and so on. Then the sequence $p_2,p_3,\ldots$ converges to p_1. Let *f* be a fleeing property and let us suppose that of the existence of its critical number κ_f the absurdity of the absurdity has been established. Let $r_\nu = p_\nu$ for each down-number ν of *f*, and $r_\nu = p_{\kappa_f}$ for each up-number ν of *f*. Then the sequence $r_1,r_2,\ldots$ converges to a sharp real number $r > p_1$; but no p_ν, and hence no element of *F*, can be indicated which lies between p_1 and *r*. Thus the intuitionist reduced continuum cannot be ordinally separable-in-itself as long as we cannot exclude fleeing properties with absurd absurdity of existence of a critical number.

Again let us suppose that G is a completely ordered fundamental sequence, demonstrating the separability-in-itself of the intuitionist full continuum. Let q_1 and q_2 be two arbitrary elements of G, such that $q_1 < q_2$ according to the virtual order of the full continuum. Let q_3 be the first element of the fundamental sequence G which lies between q_1 and q_2 according to the virtual order of the full continuum. Let q_4 be the first element of the fundamental sequence G which lies between q_1 and q_3 according to the virtual order of the full continuum, and so on. Then the sequence $q_2, q_3, \ldots$ converges to q_1. Let α be an assertion for which no method of testing is known. Choose $s_\nu = q_\nu$ as long as neither the absurdity of the absurdity nor the absurdity of α has been established, and $s_\nu = q_m$ for $\nu \geqq m$ if between the choice of s_{m-1} and that of s_m either the absurdity of the absurdity or the absurdity of α has been established. Then the sequence $s_1, s_2, \ldots$ converges to a real number $s > q_1$; but no q_ν, and hence no element of G, can be indicated which lies between q_1 and s. Thus the intuitionist full continuum cannot be ordinally separable-in-itself as long as there exist mathematical assertions for which no method of testing is known.

We adapt the definition of ordinal separability-in-itself to intuitionism in the following way: denote by $k(ab)$ the complementary species in S of the open interval ab, by $k_1(ab)$ the subspecies of $k(ab)$ whose elements are $\leqq a$ and $\leqq b$, by $k_2(ab)$ the subspecies of $k(ab)$ whose elements are $\geqq a$ and $\geqq b$. If $k(ab)$ coincides with the union of $k_1(ab)$ and $k_2(ab)$, a and b will be called *sharply different* and the interval ab will be called *substantial*. In this case, because a belongs to $k(ab)$, a belongs either to $k_1(ab)$ or to $k_2(ab)$. Hence we have either $a \leqq b$ or $a \geqq b$. Since a and b were assumed different, it follows that either $a > b$ or $a < b$. Let $a < b$ (the case $a > b$ is treated in the same way), then all elements of $k_1(ab)$ are $\leqq a$ and all elements of $k_2(ab)$ are $\geqq b$. The species $k_1(ab)$ and $k_2(ab)$ are what we have defined as conjugate removable subspecies of $k(ab)$. Furthermore, for every $a<b$ in a virtually ordered species S, the closed interval ab is the intersection of the species $x \leqq b$ and $\geqq a$ (i.e. the species whose elements are $\leqq b$ and the species whose elements are $\geqq a$), the open interval ab is the intersection of the species $\leqq b$, $\geqq a$, $\neq b$ and $\neq a$, i.e. of the species $<b$ and $>a$, and $k(ab)$ contains the union of the species $\geqq b$ and $\leqq a$. Now let S be the full intuitionist continuum. Let $\alpha(k'_1,k'_2,\ldots)$ be a point of a and $\beta(k''_1,k''_2,\ldots)$ a point of b, where $a<b$, a and b are sharply different, and each k'_ν and k''_ν are $\lambda^{(4\nu+1)}$-intervals. Let, as before, the interval k'''_ν be one of the intervals k'_ν and k''_ν, and k^{IV}_ν the other, in such a way that the centre of k^{IV}_ν never lies to the left of the centre of k'''_ν. Then, on account of $a<b$, the infinite sequence $\gamma(k'''_1,k'''_2,\ldots)$ will also be a point of a, and the infinite sequence $\delta(k^{IV}_1,k^{IV}_2,\ldots)$ will be a point of b; and for each n the centre of k^{IV}_n will either lie to the right of the centre of k'''_n or coincide with the centre of k'''_n.

We now consider the cluster (the term being used in a slightly wider sense) generated by the sequences $k^V_1,k^V_2,\ldots$, where: (i) for k^V_1 both k'''_1 and k^{IV}_1 can be chosen; (ii) for each $k^V_{\nu+1}$ both $k'''_{\nu+1}$ and $k^{IV}_{\nu+1}$ can be chosen as long as $k'''_{\nu+1}$ and $k^{IV}_{\nu+1}$ overlap; (iii) otherwise after k'''_ν only $k'''_{\nu+1}$, and after k^{IV}_ν only $k^{IV}_{\nu+1}$, can be chosen. Each sequence $k^V_1,k^V_2,\ldots$ converges to a point core e. If, for a certain m, k'''_{m+1} and k^{IV}_{m+1} do not overlap, then each e either coincides with a or coincides with b. So, if any e were to lie *between* a and b (i.e. to belong to the open interval ab) each k'''_{m+1} would overlap k^{IV}_{m+1}, and a and b would coincide. So, because coincidence of a and b is absurd, no e can lie between a and b, i.e. each e belongs to $k(ab)$. Hence, since a and b are sharply different, each e must either belong to $k_1(ab)$ and deviate from $k_2(ab)$ or belong to $k_2(ab)$ and deviate from $k_1(ab)$. But the decision between these two mutually exclusive properties *for all point cores e* is not possible, as long as we do not know a natural number m such that k'''_{m+1} and k^{IV}_{m+1} do not overlap, i.e. as long as we have not established that $a \lessdot b$. So that we have proved that, if a and b are two sharply different point cores of the intuitionist full continuum, we have either $a \lessdot b$ or $a \gtrdot b$. Thus, let us call a virtually ordered species S *ordinally separable-in-itself in the wider sense*, if in S a completely ordered fundamental sequence F can be indicated such that between any two sharply different elements of S there is an element of F. Then, on the one hand, in classical mathematics this definition is equivalent with that of ordinal separability-in-itself; on the other hand at least the intuitionist full continuum is ordinally separable-in-itself in the wider sense.

[THE ORDINAL CONNECTEDNESS] To verify the property of the ordinal connectedness for the intuitionistic continuum, we first have to specify the intuitionist sense of the word 'subdivision', and to distinguish a weak, a medium and a strong ordinal connectedness, according to whether we understand by a 'subdivision' a split, a direct composition, or a general composition. Then for the intuitionist full continuum the weak ordinal connectedness is fulfilled, but void, because, as we shall see in Chapter 5, the intuitionist full continuum does not split. And for the reduced intuitionist continuum the property is also void, because the intuitionist reduced continuum certainly does not split into a left subspecies α and a right subspecies β, such that α has a last element e and β has no first element. This follows from the following reasoning: each element of the intuitionist reduced continuum which is $>e$ cannot belong to α and therefore belongs to β; conversely, every element of β is $>e$, Similarly, α would be identical to the species of the elements of the reduced continuum which are $\leqq e$. It would follow that every element of the reduced continuum would either be $>e$ or $\leqq e$, which is not the case.

As for the medium and strong ordinal connectedness, let us consider a fleeing property f, a subspecies α of the reduced (or full) continuum C, and a subspecies β of C, consisting of all elements of C which cannot possibly belong to α. The species α is to consist of all the elements of $C<1$, and moreover of all the elements <2 from the moment that the absurdity of the absurdity of the existence of κ_f has been proved. Conversely, the species γ consisting of all elements of C which cannot possibly belong to β is easily seen to be identical with α.

Then we have the following schema:

	To α belong and only belong those elements of C which are	To β belong and only belong the elements of C which are	To γ belong and only belong the elements of C which are
So far	<1 and perhaps later those <2	$\geqq 2$ and perhaps later those $\geqq 1$	<1 and perhaps later those <2
after a proof of absurdity of κ_f's existence	only those <1	those $\geqq 2$ and those $\geqq 1$	only those <1
after a proof of abs abs of κ_f's existence	those <1 and those <2	only those $\geqq 2$	those <1 and those <2

So C is directly composed of α and β, but neither α possesses a last, nor β a first, element. Thus the strong, as well as the medium, ordinal connectedness holds neither for the intuitionist reduced continuum nor for the intuitionist full continuum.

To adapt the definition of ordinal connectedness to intuitionism, we introduce the notion of an *exhaustive sectional subdivision* of the virtually ordered species S as a subdivision into a left subspecies α and a right subspecies β, where: (i) S is directly composed of α and β; (ii) for any pair of sharply different elements a and b ($a<b$) either all elements $\leqq a$ belong to α or all elements $\geqq b$ belong to β. We see at once that if S is the full or the reduced continuum, an exhaustive sectional subdivision of S yields an intercalation subdivision of each skeleton S, since any two different elements of the skeleton are sharply different elements in the full as well as in the reduced continuum. We shall call a virtually ordered species S *completely ordinally connected* if each exhaustive sectional subdivision ε of S into a left subspecies α_ε and a right subspecies β_ε yields a *cut element* s_ε,

i.e. an element s_ε of S such that every element $<s_\varepsilon$ belongs to α and every element $>s_\varepsilon$ belongs to β. And S will be called *rigidly ordinally connected* if each exhaustive sectional subdivision of S that is predeterminate (thus in the case of the continuum yields a predeterminate intercalation subdivision in each skeleton) yields a cut element. In this way we obtain complete ordinal connectedness for the full continuum and rigid ordinal connectedness for the reduced continuum. Indeed, the intercalation subdivision d_ε corresponding to the exhaustive sectional subdivision ε of the intuitionist continuum yields a real number t_ε, such that every real number $\lessdot t_\varepsilon$ belongs to α and every real number $\gtrdot t_\varepsilon$ belongs to β. Let r be a real number $<t_\varepsilon$. If r belonged to β we should have to state the absurdity of r belonging to α, thus in particular of $r \lessdot t_\varepsilon$, so that the relation $r \geqq t_\varepsilon$ would hold, which would clash with our supposition $r<t_\varepsilon$. The absurdity of r belonging to β, thus proved, entails that r belongs to α, and in the same way we define that every real number $>t_\varepsilon$ belongs to β, so that t_ε proves to be a cut element with regard to ε.

[THE ORDINAL EVERYWHERE-DENSITY] As the ordinal everywhere-density includes only a part of the ordinal separability-in-itself, there is a chance that it will maintain its validity for the intuitionist continuum in its classical form, after the classical sense of 'between' has been extended in the sense explained above under the investigation of ordinal density-in-itself. That this is indeed the case can be proved by deducing a point $p(h_1,h_2,\ldots)$ from two points arbitrarily given, $p'(h'_1,h'_2,\ldots)$ and $p''(h''_1,h''_2,\ldots)$, where h_ν, h'_ν and h''_ν denote λ-intervals, not necessarily $\lambda^{(4\nu+1)}$-intervals, in the following way. If h'_ν and h''_ν overlap, then $h_\nu = h'_\nu$. If h'_ν and h''_ν overlap, but $h'_{\nu+1}$ and $h''_{\nu+1}$ do not overlap (so that $h'_{\nu+2}$ and $h''_{\nu+2}$ lie apart), then there certainly is an intersection interval τ of the intersection interval ρ of h'_ν and h''_ν and the intervening interval σ of $h'_{\nu+2}$ and $h''_{\nu+2}$. For if, for example, σ were to lie entirely outside h'_ν, neither $h'_{\nu+2}$ nor $h''_{\nu+2}$ would lie within h'_ν. Thus we can define $h_{\nu+1}$ as the greatest (and in doubtful cases, amongst the greatest, the leftmost) λ-

interval lying within this interval τ, so that $h_{\nu+1}$ lies firstly, as required, within h_ν, and secondly apart from and between $h'_{\nu+2}$ and $h''_{\nu+2}$. Furthermore, for $\nu > n+1$, h_ν is defined as concentric with $h_{\nu-1}$ and half the size of $h_{\nu-1}$.

We shall now suppose that p' and p'' deviate. Then also

(i) *p and p' deviate.*

For, if p and p' coincide, then in the above construction h_ν and h'_ν would have to overlap for every ν, so h'_ν and h''_ν would have to overlap for every ν and p' and p'' would coincide. In the same way we see that

(ii) *p and p'' deviate.*

Furthermore, if both $p \mathrel{\circ\!\!>} p'$ and $p \mathrel{\circ\!\!>} p'$ were impossible, then again in the above construction h'_ν and h''_ν would have to overlap for every ν and again p' and p'' would coincide. So we have the impossibility of the combination of $p \mathrel{\circ\!\!\leqq} p'$ and $p \mathrel{\circ\!\!\leqq} p''$, therefore the impossibility of the combination of $p \leqq p'$ and $p \leqq p''$. Thus also

(iii) *the combination of $p < p'$ and $p < p''$ is impossible.*

In the same way we see that

(iv) *the combination of $p > p'$ and $p > p''$ is impossible.*

From (i)–(iv) it follows that *p lies between p' and p''.*

The above construction of p from p' and p'' can be carried out for the reduced as well for the full continuum, so that the ordinal everywhere-density has been re-established for both.

[THE ORDINAL LOCAL COMPACTNESS] That neither the intuitionist reduced continuum nor the intuitionist full continuum is ordinally locally compact we prove by constructing the following *predeterminate* infinite sequence $I_1, I_2, \ldots$ of closed intervals, each of which is a *proper* subspecies of the immediately preceding one, and not possessing the required property.

Let f be a fleeing property, two-sided with regard to parity. For a down-number ν of f, I_ν will have the end elements $-\frac{1}{2}-2^{-\nu}$ and $\frac{1}{2}+2^{-\nu}$; for an up-number ν of f, I_ν will have the end elements $-\frac{1}{2}-2^{-\nu}$ and $-\frac{1}{2}+2^{-\nu}$ if κ_f is odd, but $\frac{1}{2}-2^{-\nu}$ and $\frac{1}{2}+2^{-\nu}$ if κ_f is even. Then neither the intuitionist reduced continuum nor the intuitionist full continuum possesses an element common to all I_ν.

To adapt the definition of ordinal local compactness to intuitionism, we shall say that a telescopic infinite sequence of closed intervals $I_1,I_2,\ldots$ of a virtually ordered species S *contracts completely*, or is a *sequence of complete contraction*, if for each substantial interval I of S a ν can be indicated such that I cannot possibly be a subspecies of I_ν. Furthermore we shall say that a telescopic infinite sequence of closed intervals $I_1,I_2,\ldots$ of S is a *hollow sequence*, if for each element e of S a ν can be indicated such that e cannot possibly belong to I_ν.

Let us consider a completely contracting telescopic infinite sequence of closed intervals $I_1,I_2,\ldots$ of the intuitionist full continuum C, and let its intersection be denoted by I. Let each I_ν be the intersection of the convergent telescopic infinite sequence of binary closed intervals ${}_\nu j_1, {}_\nu j_2, \ldots$ If we replace each ${}_\nu j_\eta$ by ${}_\nu i_\eta = \cap_{\mu \leqq \nu}\, {}_\mu j_\eta$, then because, for each ν, $I_\nu = \cap_{i \leqq \nu} I_i$, each I_ν is also identical to the intersection of the convergent telescopic infinite sequence of binary closed intervals ${}_\nu j_1, {}_\nu j_2, \ldots$ If we replace each ${}_\nu j_\mu$ by ${}_\nu i_\mu = \cap_{i \leqq \nu}\, {}_i j_\mu$, then because, for each $\nu, I_\nu = \cap_{i \leqq \nu} I_i$, each I_ν is also identical to the intersection of the convergent telescopic infinite sequence of binary closed intervals ${}_\nu i_1, {}_\nu i_2, \ldots$, each ${}_{\nu+1} i_\mu$ being a subspecies of ${}_\nu i_\mu$. Let u and v be two integers such that each ${}_\nu i_\mu$ lies between u and v. Let U be the closed interval uv, and let W_ν be the species of the closed intervals which each have the end points of a $\lambda^{(4\nu+1)}$-interval as their end points and which overlap U. Let for each natural number σ n_σ be a natural number such that no element of W_σ can be a subspecies of I_{n_σ}. We may suppose that, for each σ, $n_{\sigma+1} \geqq n_\sigma$.

Furthermore let, for each natural number σ, m_σ be a natural number such that the size of ${}_{n_\sigma}i_{m_\sigma}$ is less than $2^{-4\sigma}$. We may suppose that for each σ $m_{\sigma+1} \geqq m_\sigma$. Then, for each σ, ${}_{n_{\sigma+1}}i_{m_{\sigma+1}}$ is a subspecies of ${}_{n_\sigma}i_{m_\sigma}$, and the intersection $\cap_{\mu \geqq 1}\, {}_{n_\mu}i_{m_\mu}$, which we shall denote by I_0, is a single point core. Now we may write, for each σ,

$$I_{n_\sigma} = \bigcap_{\mu \geqq 0} {}_{n_\sigma}i_{m_{\sigma+\mu}},$$

$$I_0 = \bigcap_{\mu \geqq 0} {}_{n_{\sigma+\mu}}i_{m_{\sigma+\mu}},$$

where ${}_{n_{\sigma+\mu}}i_{m_{\sigma+\mu}} \subset {}_{n_\sigma}i_{m_{\sigma+\mu}}$, so that $I_0 \subset I_{n_\sigma}$ for each σ, whence $I_0 \subset \cap_{\nu \geqq 1} I_{n_\nu}$, and $I_0 \subset I$. On the other hand, we may write, for each σ,

$$I_0 = \bigcap_{\nu \geqq 1} {}_{n_\nu}i_{m_\nu}$$

$$I = \bigcap_{\nu \geqq 1} I_{n_\nu},$$

where $I_{n_\nu} \subset {}_{n_\nu}i_{m_\nu}$, so that $I \subset I_0$. Hence we have $I = I_0$, and I also proves to be a single point core. If the end elements of I_ν are sharp point cores for each ν and if, moreover, the sequence $I_1, I_2, \ldots$ is a predeterminate one, then I will also be a sharp point core. Hence, if we call a virtually ordered species S *completely locally compact*, if each sequence of complete contraction possesses one and only one intersection element, and *rigidly locally compact* if each predeterminate sequence of complete contraction possesses one and only one intersection element, then the intuitionist full continuum is completely locally compact, and the intuitionist reduced continuum is rigidly locally compact.

Let us suppose that in the intuitionist full continuum a hollow sequence $I_1, I_2, \ldots$ could exist. Then for each natural number ν there would exist an n_ν such that for no integer could an I_{n_ν} contain the binary fraction $a \cdot 2^{-\nu}$. Hence the sequence would be of complete contraction, and possess one and only one intersection point core. In the case that the end elements of each I_ν are sharp point cores and elements of each I are sharp point cores and the sequence $I_1,I_2,\ldots$ is predeterminate, the intersection point core of the sequence would also be sharp. So we see that *in the intuitionist full continuum no hollow sequence, and in the intuitionist reduced continuum no predeterminate hollow sequence, can exist.*

5

The bunch theorem[39]

Let us consider a spread key σ, and let t be an arbitrarily chosen node of order 1. Let B be a crude block of $K(\sigma)$. Then there exists a mathematical deduction h_t starting from the indication of B and proving that t is *blocked*, i.e. that no arrow containing this node can avoid B.

For this mathematical deduction no data are available other than the species B and the system of constructional relations between the nodes of σ. Now the latter can be split up into such 'basic relations' as exist between an immediate predecessor and an immediate descendant or between two contiguous elements of a row of nodes, and such cases of partial lacking the aforesaid basic relations as are represented by the 'basic properties' of being a node of order 1 or of being the first or the last element of a row of nodes. Consequently, if we consider h_t as a pseudo well-ordered species k_t built up of basic facts d and immediate inferences e, both reduced to as small atoms as possible, this can only be made in such a way that an e, stating that the element $\nu_1 \ldots \nu_\rho$ of the stump τ carried by B is blocked, deduces this (neither $\nu_1 \ldots \nu_{\rho-1}\nu_\rho - 1$ nor $\nu_1 \ldots \nu_{\rho-1}\nu_\rho + 1$ having anything to do with the free arrows passing through $\nu_1 \ldots \nu_{\rho-1}\nu_\rho$) either from the knowledge that $\nu_1 \ldots \nu_{\rho-1}$ is blocked or from the knowledge that $\nu_1 \ldots \nu_\rho \nu$ is blocked for each ν. In the former case we shall call e a *ζ-inference*, in the latter case a *F-inference.*

Let g_t be the species of those nodes of τ whose property of being blocked is used by k_t after having been stated previously to k_t, or is deduced in the course of k_t. Then the immediate inferences e of k_t are *conservative*, i.e. in case each element of g_t, whose property of being blocked is used by e after having been stated previously to e, dominates a well-ordered stump sector carried by a subspecies of B and belonging to g_t, then an element of g_t whose property of being blocked is deduced by e likewise dominates a well-ordered stump sector carried by a subspecies of B and belonging to g_t. Moreover each element of g_t, whose property of being blocked is stated previously to k_t, dominates a well-ordered stump sector carried by a subspecies of B (consisting in this case of a single node of B). So the inductive method leads to the result that each node of g_t possesses the property in question and that in particular the node t itself dominates a well-ordered stump sector τ_t carried by a subspecies B_t of B. But then $\tau_\omega = \Sigma_t \tau_t$ is a well-ordered stump carried by the well-ordered block $B_\omega = \Sigma_t B_t$ of $K(\sigma)$, which is a subspecies of B. Hence we have proved that *every crude spread key block contains a well-ordered block* (so that in particular every proper spread key block is a well-ordered block).

By an analogous reasoning we deduce that *every completed crude spread clue block contains a well-ordered completed block*, and that *every crude spread clue block contains a pseudo well-ordered block.*

As a particular case of the latter theorem let us consider a bunch σ, a crude block B of σ, the pseudo well-ordered block B_ω of σ derived from B, its completion B°_ω and the completed stump τ°_ω carried by B°_ω. In this particular case all rows of nodes of σ are finite. So in the marked well-ordered species F corresponding to τ°_ω, each generating operation having the passage from a finite row of nodes to its dominant as its image in τ°_ω must be of the first kind. Hence F is a bounded well-ordered species. But then F is finite, and the same holds for B°_ω, for τ°_ω and for B_ω. So we see that *for a bunch every crude block contains a finite block.*

A corollary of this theorem states that *in every crude block of a bunch a block is contained for which a natural number can be calculated bounding the order of its nodes.*

Again a particular case of this corollary is the

BUNCH THEOREM *If to each free arrow e of a bunch σ a natural number n_e has been assigned, then a natural number m can be calculated such that each n_e is determined by the node of order m belonging to e.*

As a first application of the bunch theorem (using only its specialization for a cluster) let us consider a *full function* $y=f(x)$ of the *unity continuum* U, i.e. a function which assigns a real number y to each real number x contained in the $\lambda^{(0)}$-interval U with centre $\frac{1}{2}$. Then a cluster point spread $S(U)$ consisting of the 'quartering' points $p(k'_{\nu_1}, k''_{\nu_2},\ldots)$, each $k^{(\nu)}$ being a $\lambda^{(4\nu+1)}$-interval overlapping U, coincides with U. Let $\sigma(U)$ be a strictly individualized cluster generating $S(U)$. Then by $y=f(x)$ to each point of $S(U)$ (or, which amounts to the same, to each arrow of $\sigma(U)$) a one-dimensional Cartesian point $\pi(\lambda_1,\lambda_2,\lambda_3,\ldots)$ has been assigned in such a way that coincident one-dimensional Cartesian points correspond to coincident points of $S(U)$.

Under these circumstances, by virtue of the bunch theorem, for each natural number n a natural number $m=\psi(n)$ can be calculated such that for each arrow e of $\sigma(U)$ the λ_n of the point π assigned to e is determined by the node of order m belonging to e. Consequently the species of these λ_ns is finite. Hence for each n an integer $\chi(n)$ can be indicated such that the maximum size of the λ_ns is $2^{-\chi(n)}$, where $\chi(n+1)\geqq\chi(n)+1$, so that $\chi(n)$ increases indefinitely (i).

Let, for each $k^{(\nu)}$, $^{\circ}k^{(\nu)}$ be an interval concentric with, and three-quarters of the size of, $k^{(\nu)}$. Then two real numbers x_1 and x_2 of U, whose distance is $\lessdot 2^{-4\nu-3}$, always lie together within one and the same suitable chosen $^{\circ}k^{(\nu)}$, so that a point p_1 of x_1 and a point p_2 of x_2 can be indicated which both belong to $S(U)$ and are generated by arrows of $\sigma(U)$ having equal νth nodes (ii).

Let n_1 be a natural number, and let us find sufficient conditions under which, by the full function $y=f(x)$, two real numbers x_1 and x_2 of U get assigned real numbers y_1 and y_2 differing by less than 2^{-n_1}.

By virtue of (i), firstly a natural number $n_2=\varphi_1(n_1)$ exists such that each λ_{n_2} is smaller than 2^{-n_1}; secondly a natural number $n_3=\varphi_2(n_2)$ exists such that the λ_{n_2}, assigned to any arrow e of $\sigma(U)$, depends only on the node of order n_3 belonging to e, so that to two points of $S(U)$ generated by arrows of $\sigma(U)$ with equal nodes of order n_3 the same λ_{n_2} has been assigned.

Furthermore, by virtue of (ii), a natural number $n_4=\varphi_3(n_3)$ exists such that two real numbers x_1 and x_2 of U, whose difference is $\lessdot 2^{-n_\nu}$, contains points p_1 and p_2 respectively, both belonging to $S(U)$ and generated by arrows of $\sigma(U)$ having equal nodes of order n_3.

Consequently the full function $y=f(x)$ assigns to p_1 and p_2 two points with the same λ_{n_2} and in this way assigns to x_1 and x_2 two real numbers y_1 and y_2 differing by less than 2^{-n_1}.

So, an arbitrary full function $y=f(x)$ being given, for each natural number n_1 a natural number $n_4=\varphi(n_1)$ can be calculated such that the given function assigns two real numbers y_1 and y_2, differing by less than 2^{-n_1}, to each pair of real numbers x_1 and x_2, whose difference is less than 2^{-n_4}.

Altogether we have proved:

Each full function of the unity continuum is uniformly continuous.

The proof of this fundamental theorem essentially relies upon well-ordered species and their close relation to spread keys. But there is a weaker form of the theorem (equivalent according to classical mathematics), expressing the *impossibility of the existence of a point core of discontinuity for a full function.* It is in this form that the theorem can be deduced as an immediate consequence of the fundamental thoughts of intuitionism without using spread keys or well-ordered species.

For, let us suppose that $y=f(x)$ is a full function of U; ξ_0 a real number belonging to U; $\xi_1,\xi_2,\ldots$ an infinite sequence of real numbers of U converging to ξ_0; t a natural number; and that $|f(\xi_\nu)-f(\xi_0)|>1/t$ for every ν.

Let g be a fleeing property and κ_g its critical number. We define an infinite sequence of real numbers $q_1,q_2,\ldots$ in the following way: $q_\nu=\xi_\nu$ for $\nu\leqq\kappa_g$ and $q_\nu=\xi_{\kappa_g}$ for $\nu\geqq\kappa_g$. This sequence converges to a real number q_0, to which no real number $f(q_0)$ can be assigned.

In much the same way as we deduced the uniform continuity of full functions of the unity continuum, we show that if an infinite sequence $f_1(x),f_2(x),\ldots$ of uniformly continuous functions of the unity continuum converges everywhere to a limiting function $f(x)$, then not only is $f(x)$ again a uniformly continuous function, but moreover the convergence of the $f_\nu(x)$ to $f(x)$ is uniform. Consequently intuitionism cannot apply Baire's method which, starting from the class of uniformly continuous functions, introduces new classes of functions of U ever and again by means of convergent infinite sequences of functions previously acquired. But nevertheless intuitionism, no less than classical mathematics, is led to an extension of the domain of functions of a real variable beyond the class of uniformly continuous functions. This extension is brought about in a natural way by the intuitionistic theory of integration, whose natural field of application consists of the *measurable functions*. A measurable function is not a full function, but it is 'almost full', and is moreover 'almost everywhere' the limiting function of an infinite sequence of full functions. Furthermore, if an infinite sequence $f_1(x),f_2(x),\ldots$ of measurable functions converges 'almost everywhere' to a function $f(x)$, then $f(x)$ is likewise measurable. For the rest several theorems on intuitionistic measurable functions are worded much like well-known theorems of classical mathematics, but have a different meaning and require an entirely different deduction.

As a second application of the bunch theorem we shall consider the Heine–Borel theorem, which in intuitionistic as well as in classical mathematics is continually used in the theory of functions of a real variable. For the sake of simplicity we shall confine our attention to the Cartesian plane C.

Let P be a point core of C. Let us call a point core species of C containing each point core at a distance $\leqq 2^{-m}$ from P an *m-neighbourhood* of P, and a point core species which is an m-neighbourhood of P for some m a *neighbourhood* of P. Let us call a point core species Q *consolidated* if it is bounded, and to each convergent infinite sequence d of different point cores of Q there exists a point of accumulation of d likewise belonging to Q. Then for the Cartesian plane the classical form of the Heine–Borel theorem runs as follows:

If to each point core P of a consolidated point core species Q of the Cartesian plane a neighbourhood has been made to correspond, then from the species O of these neighbourhoods a finite subspecies O_1 can be selected such that for each point core P of Q at least one element of O_1 is a neighbourhood.

In this form the theorem certainly does not hold for intuitionism, as we see by the following two examples, both starting from a fleeing property f and its critical number κ_f.

(α) Let Q be the consolidated point core species which is the union of the sequence of point cores P_ν ($\nu=1,2,\ldots$), each point core P_ν having coordinates $(c_\nu,0)$, where $c_\nu=2^{-\nu}$ for $\nu\leqq\kappa_\nu$, and $c_\nu=2^{-\kappa_f}$ for $\nu\geqq\kappa_f$, and the species of the point cores of accumulation of this sequence. To each point core P_ν of the sequence we assign the λ-square with P_ν as its centre and with side length $\frac{1}{2}c_\nu$; to each point core of accumulation of the sequence we assign the λ-square with the origin as its centre and with side length 1.

(β) Let Q be the consolidated point core species consisting of the sequence of point cores P_ν ($\nu=1,2,\ldots$), each point core P_ν having the coordinates $(c_\nu,0)$, where $c_\nu=1$ for $\nu\leqq\kappa_f$ and $c_\nu=0$ for $\nu>\kappa_f$. To each point core P_ν of the sequence we assign the λ-square with P_ν as its centre and with side length 1.

In both examples the conditions required by the classical Heine–Borel theorem are satisfied, but the assertion of the theorem defaults. For, in both examples there is no guarantee that the origin belongs to Q, so that the λ-interval of side length 1 is not available as an element of O_1.

To explain this failure of the classical theorem let us examine a representative form of its classical proof, in which five stages can be distinguished:

(1) *The P-diameter of a neighbourhood ω of P is defined as the upper bound of the diameters of the circles which can be described with P as centre such that all their interior points belong to ω.* For intuitionism this definition is illusory, as is shown by the following example. Let f be an arbitrary fleeing property, let P be the origin, and let us choose for ω the union $\cup_{\nu \geqq 1}\lambda_\nu$, where each λ-square λ_ν has the origin as its centre, while its side length is 2 if ν possesses the property f, and 1 if ν does not possess this property.

(2) *The subspecies O(P) of O is introduced as the species of those elements of O which are neighbourhoods of P, and the span of P is defined as a real number $\alpha_P \gtrdot 0$, which is the upper bound of the P-diameters of the elements of O(P).* For intuitionism the latter definition not only naturally shares the illusive character of the P-diameter, but even if for each element of $O(P)$ the P-diameter could be determined, there would not necessarily exist an α_P. This is demonstrated by the species of the λ_ν of the preceding paragraph, and is also exemplified in the case that for a point core P' it is doubtful if it belongs to Q, but certain that *if* it belongs to Q, a neighbourhood of large P-diameter will have to be assigned to it. See the above example (α).

(3) *In order to prove that α_P cannot take indefinitely small values, the supposition is introduced that an infinite sequence of point cores $P_1, P_2, \ldots$ exists whose span converges to zero. It is then argued that the obvious existence of a convergent infinite*

subsequence $P_{\rho_1}, P_{\rho_2}, \ldots$ whose span converges to zero would be a consequence of this supposition. By intuitionism this consequence is not recognized as, for example (κ_f being the critical number of the fleeing property f), the infinite sequence $(c_1,0),(c_2,0),\ldots$, where $c_\nu = 1+2^{-\nu}$ for $\nu < \kappa_f$ and $c_\nu = 2^{-\nu}$ for $\nu \geqq \kappa_f$, does not contain a convergent infinite subsequence.

(4) *From the absurdity of the existence of indefinitely small values of α_P the existence of a natural number n is inferred such that $\alpha_P > 2^{-n}$ for each P.* Again this inference cannot be recognized by intuitionism, as is elucidated by the following example. Let α be a mathematical assertion such that no possibility of testing is known. Let $c_1, c_2, \ldots$ be a freely proceeding sequence of decreasing positive rational numbers such that $2^{-\nu} < c_\nu < 2^{-\nu+1}$, as long as at the choice of c_ν neither the truth nor the absurdity of α has been established, but $c_m = 2^{-m+1}$ and $c_{m+\nu} = 2^{-m} + 2^{-m-\nu}$ $(\nu = 1,2,\ldots)$ if between the choice of c_{m-1} and c_m either the truth or the absurdity of α has been established.

(5) *In each $k^{(n+2)}$-square which contains at least one point core of Q we choose a point core of Q. In this way we determine a finite species $P_1, P_2, \ldots, P_r$ of point cores of Q, and to each P_ν we assign an element $o(P_\nu)$ of $O(P_\nu)$ with P-diameter $\gtrdot 2^{-n-1} \cdot \sqrt{2}$. Then $o(P_1), o(P_2), \ldots, o(P_r)$ constitute the finite species of neighbourhoods O_1 required by the Heine–Borel theorem.*

Here intuitionism makes the objection that a $k^{(n+2)}$-square k_1 may exist which has neither been proved to contain nor not to contain a point core of Q. This may leave the possibility that at some time a point core of Q contained in k_1 will be found for which none of the elements of O_1, determined in the way indicated above, will furnish a neighbourhood. See the above example (β).

But all the above objections are cleared away by confining the theorem to compact located point core species, whose definition was given, and whose coincidence with bunch point spreads was proved, in Chapter 2. It is this coincidence that together with the bunch theorem enables us to deduce the Heine–Borel theorem in its intuitionistic form as follows (for the Cartesian plane).

If to each point core P of a compact located point core species Q of the Cartesian plane a neighbourhood of P has been made to correspond, then from the species O of these neighbourhoods a finite subspecies O_1 can be selected such that for each point core P of Q at least one element of O_1 is a neighbourhood.

For, let $\sigma(Q)$ be a strictly individualized bunch generating a bunch point spread $s(Q)$ coinciding with Q. Then to each point of $s(Q)$, so to each free arrow of $\sigma(Q)$, a neighbourhood and therefore a natural number m characterizing this neighbourhood has been assigned. Then, by virtue of the bunch theorem, a natural number r can be calculated such that, for each free arrow of $\sigma(Q)$, the natural number m assigned to it is determined by its node of order r, so it can only take a finite number of values. If c is the greatest of these values, then to each free arrow of $\sigma(Q)$, so to each point core P of Q, a c-neighbourhood $o(P)$ has been assigned.

Now from the species M of the point cores $(a\cdot 2^{-c-1}, b\cdot 2^{-c-1})$ (a and b arbitrary integers) of the Cartesian plane we select a finite subspecies N $(=\{\pi_1,\pi_2,\ldots,\pi_k\})$ in such a way that each point core of M having a distance $\leqq \frac{3}{4}\cdot 2^{-c-1}$ from Q is taken into N, and each point core of M having a distance $\geqq \frac{7}{8}\cdot 2^{-c-1}$ from Q is not taken into N. Then to each π_ν a point core $P(\pi_\nu)$ of Q can be assigned, from which it has a distance $\leqq 2^{-c-1}$. Furthermore every point core of Q has a distance $<\frac{3}{4}\cdot 2^{-c-1}$ from one of the k point cores π_ν, and so a distance $<\frac{7}{4}\cdot 2^{-c-1}$ from one of the k point cores $P(\pi_\nu)$ of Q. So every point core of Q belongs to one of the k neighbourhoods $o(P(\pi_\nu))$.

As a third application of the bunch theorem (using only its specialization for a cluster) we shall prove that '*the continuum does not split*'. We understand by this short expression that *the continuum cannot contain a removable proper subspecies*, or, which amounts to the same, that *if the continuum splits into two conjugate removable subspecies, one of these subspecies is equal to the continuum and the other one is empty.* Evidently it will be sufficient to deduce this theorem for the unit continuum.

Let $s(U)$ be the cluster point spread coincident with U consisting of the 'quartering' points $p(k'_{\nu 1}, k''_{\nu 2}, \ldots)$, each $k^{(\nu)}$ being a $\lambda^{(4\nu+1)}$-interval overlapping U. Let $\sigma(U)$ be a strictly individualized cluster generating $s(U)$, and $b(U)$ the corresponding blank cluster spread. Let us suppose that U_1 and U_2 are conjugate removable subspecies of $s(U)$, and b_1 and b_2 the corresponding conjugate removable subspecies of $b(U)$. By this splitting up of $b(U)$ an assignment either of the natural number 1 or of the natural number 2 to each free arrow of $\sigma(U)$ is determined. So, by virtue of the bunch theorem, a natural number m and two conjugate removable subspecies ${}_1N_m$ and ${}_2N_m$ of the species N_m of the nodes of order m of $\sigma(U)$ can be indicated in such a way that $\sigma(U)$ assigns a point of s_1 and U_1 to each of its arrows containing an element of ${}_1N_m$, and a point of s_2 and U_2 to each of its arrows containing an element of ${}_2N_m$.

Consequently, the points which $\sigma(U)$ assigns to an arrow containing an element of ${}_1N_m$ and to an arrow containing an element of ${}_2N_m$ can never coincide.

Let us suppose that an element of ${}_1N_m$ as well as an element of ${}_2N_m$ can be indicated. An element θ_1 of ${}_1N_m$ and an element θ_2 of ${}_2N_m$ can then be indicated such that the $k^{(m)}$ assigned by $\sigma(U)$ to θ_1 and the $k^{(m)}$ assigned by $\sigma(U)$ to θ_2 are identical or overlap. In both cases an arrow containing the element θ_1 of ${}_1N_m$ and an arrow containing the element θ_2 of ${}_2N_m$ can be indicated to which $\sigma(U)$ assigns coincident points. The supposition is thus seen to lead to an absurdity. Furthermore from the impossibility of indicating an element of ${}_1N_m$ as well as an element of ${}_2N_m$ it follows that either ${}_1N_m$ or ${}_2N_m$, so also b_1 or b_2, either s_1 or s_2, and either U_1 or U_2, are empty, i.e. U does not split.

In particular the continuum does not split into the species of the rational and the species of the irrational point cores, and the assertion that each real number is either rational or irrational is contradictory. This shows that the non-contradictority of the principle of the excluded third, proved in Chapter 1 for finite species of assertions, does not extend to general species of assertions.

The non-splitting property also holds for the μ-systems introduced in Chapter 2 if they are *connected*, i.e. do not split into two μ-systems lying apart from each other. To prove this, let V be a connected μ-system, and $e(V)$ the cluster point spread coinciding with V consisting of the (n-dimensional) 'quartering' points $p(k'_{\nu_1}, k'_{\nu_2}, \ldots)$, each $k^{(\nu)}$ being a $\lambda^{(4\nu+\mu+1)}$-interval overlapping V. Let $\sigma(V)$ be a strictly individualized cluster generating $s(V)$, and $b(V)$ the corresponding blank cluster spread. Let us suppose that V_1 and V_2 are conjugate removable subspecies of V; s_1 and s_2 the corresponding conjugate removable subspecies of $s(V)$; b_1 and b_2 the corresponding removable subspecies of $b(V)$. By this splitting of $b(V)$ an assignment either of the natural number 1 or of the natural number 2 to each arrow of $\sigma(V)$ is determined. So, by virtue of the bunch theorem, a natural number m and two conjugate removable subspecies ${}_1N_m$ and ${}_2N_m$ of the species of the nodes of order m of $\sigma(V)$ can be indicated in such a way that $\sigma(V)$ assigns a point of s_1 and V_1 to each of its arrows containing an element of ${}_1N_m$, and a point of s_2 and V_2 to each of its arrows containing an element of ${}_2N_m$.

Consequently the points that $\sigma(V)$ assigns to an arrow containing an element of ${}_1N_m$ and to an arrow containing an element of ${}_2N_m$ can never coincide.

Let us suppose that an element of ${}_1N_m$ as well as an element of ${}_2N_m$ can be indicated.

Let ${}^1k^{(m)}$ be a $k^{(m)}$ assigned by $\sigma(V)$ to an element of ${}_1N_m$, and ${}^2k^{(m)}$ a $k^{(m)}$ assigned by $\sigma(V)$ to an element of ${}_2N_m$, and a ${}^0k^{(m)}$ a $k^{(m)}$ which is a k-hyperinterval belonging to V. If then a ${}^1k^{(m)}$ would exist, identical with a ${}^2k^{(m)}$, $\sigma(V)$ would assign coinciding points to an arrow containing an element of ${}_1N_m$ and to an arrow containing an element of ${}_2N_m$. This being contradictory, no ${}^1k^{(m)}$ can at the same time be a ${}^2k^{(m)}$. We shall now distinguish two cases.

First we suppose that a ${}^0k^{(m)}$ which is a ${}^1k^{(m)}$, as well as a ${}^0k^{(m)}$ which is a ${}^2k^{(m)}$, can be indicated. Then we can also indicate a hyperinterval i_1 which is a ${}^0k^{(m)}$ as well as a ${}^1k^{(m)}$, and which touches a hyperinterval i_2 which is a ${}^0k^{(m)}$ as well as a ${}^2k^{(m)}$. Let i_3 be a $k^{(m)}$ overlapping both i_1 and i_2. Then either i_1 and i_3 or i_2 and i_3 constitute a ${}^1k^{(m)}$ and a ${}^2k^{(m)}$ having a hyperinterval belonging to V in common, so that again we are led to the contradictory result that the points which $\sigma(V)$ assigns to an arrow containing an element of ${}_1N_m$ and to an arrow containing an element of ${}_2N_m$ coincide.

Next we suppose that each ${}^0k^{(m)}$ is a ${}^1k^{(m)}$. Let i_2 be a ${}^2k^{(m)}$, arbitrarily chosen. It certainly is overlapped by some ${}^0k^{(m)}$, which we shall call i_1. Then i and i_2 constitute a ${}^1k^{(m)}$ and a ${}^2k^{(m)}$ having a hyperinterval belonging to V in common, so that once more we are led to the contradictory result of the preceding paragraph. From the supposition that each ${}^0k^{(m)}$ is a ${}^2k^{(m)}$ the same contradiction is deduced.

So our supposition that an element of ${}_1N_m$ as well as an element of ${}_2N_m$ can be indicated proves to be absurd. Consequently either ${}_1N_m$ or ${}_2N_m$, so also either b_1 or b_2, either s_1 or s_2, and either V_1 or V_2, are empty, i.e. V does not split.

The non-splitting property of a connected μ-system furnishes a new and simple proof of the contradictority of any ordering of the continuum, as long as mathematical assertions are known for which no method of testing is available. This proof rests on the following lemma.

Each ordering of the continuum must necessarily be identical with, or inverse to, its intuitive virtual order.[40]

For, let ω be a hypothetical order of the continuum, and let us denote the ordering relations ω by $\dot{>}$ and $\dot{<}$. Let us consider the point cores (x,y) of the Cartesian plane belonging to the μ-

system S_μ containing those and only those $k^{(\mu)}$-squares of the unity square which satisfy the inequality $y > x$, and do not touch the straight line $y = x$. Then by virtue of the non-splitting property of S_μ for any μ, either for each μ all point cores of S_μ satisfy the relation $y \mathrel{\dot{>}} x$, or for each μ all point cores of S_μ satisfy the relation $y \mathrel{\dot{<}} x$.

In the former case (because each point core (x,y) of the unity square satisfying the relation $y \gtrdot x$ belongs to some S_μ of suitable index μ) the relation $y \gtrdot x$ implies $y \mathrel{\dot{>}} x$ in the unit continuum, and consequently the relation $y \lessdot x$ implies the relation $y \mathrel{\dot{<}} x$.

Furthermore, if for any pair of real numbers (x,y) of the unit continuum the relation $y \mathrel{\dot{>}} x$ holds, then the relation $y \mathrel{\dot{<}} x$, so the relation $y \lessdot x$, is absurd for this pair, and consequently the relation $y > x$ holds. From this it follows that for the pair of real numbers (x,y) of the unit continuum the relation $y < x$ implies $y \mathrel{\dot{<}} x$. On the other hand, if $y_0 \mathrel{\dot{<}} x_0$ holds for ω, then $y_0 < x_0$ can be added to the species of the relations $y < x$, $y > x$, $y = x$ of the intuitive virtual order of the unit continuum in a non-contradictory manner. Consequently (by virtue of the inextensibility of virtual order) this relation must belong to the intuitive virtual order of the unit continuum. So $y \mathrel{\dot{<}} x$ also implies $y < x$, and the relations $y \mathrel{\dot{<}} x$ and $y < x$ are equivalent.

In the latter case an analogous reasoning establishes the equivalence of the relations $y \mathrel{\dot{>}} x$ and $y > x$.

From the lemma thus proved, establishing the absurdity of an ordering of the continuum deviating from the intuitive ensures that any ordering of the continuum is impossible as long as mathematical assertions are known for which no method of testing is available.

Appendix

Introvert science, directed at beauty, does not carry risks for consequences.

The stock of mathematical entities is a real thing, for each person, and for humanity.

The inner experience (roughly sketched):
twoity;
twoity stored and preserved aseptically by memory;
twoity giving rise to the conception of invariable unity;
twoity and unity giving rise to the conception of unity plus unity;
threeity as twoity plus unity, and the sequence of natural numbers;
mathematical systems conceived in such a way that a unity is a mathematical system and that two mathematical systems, stored and aseptically preserved by memory, apart from each other, can be added;
etc.

Fragments from a lecture 'Changes in the relation between classical logic and mathematics. (The influence of intuitionistic mathematics on logic)', given in November 1951:

Classical logic presupposed that independently of human thought there is a *truth*, part of which is expressible by means of sentences called 'true assertions', mainly assigning certain properties to certain objects or stating that objects possessing certain properties exist or that certain phenomena behave according to certain laws. Furthermore classical logic assumed the existence of general linguistic rules allowing an automatic deduction of new true assertions from old ones, so that starting from a limited stock of 'evidently' true assertions, mainly founded on experience and called axioms, an extensive supplement to existing human knowledge would theoretically be accessible by means of linguistic operations independently of experience. Finally, using the term 'false' for the 'converse of true', classical logic assumed that in virtue of the so-called 'principle of the excluded third' each assertion, in particular each existence assertion and each assignment of a property to an object or of a behaviour to a phenomenon, is either true or false independently of human beings knowing about this falsehood or truth, so that, for example, contradictority of falsehood would imply truth whilst an assertion α which is true if the assertion β is either true or false would be universally true. The principle holds if 'true' is replaced by 'known and registered to be true', but then this classification is variable, so that to the wording of the principle we should add 'at a certain moment'.

As long as mathematics was considered as the science of space and time, it was a beloved field of activity of this classical logic, not only in the days when space and time were believed to exist independently of human experience, but still after they had been taken for innate forms of conscious exterior human experience. There continued to reign some conviction that a mathematical assertion is either false or true, whether we know it or not, and that after the extinction of humanity mathematical truths, just as laws of nature, will survive. About half a century ago this was expressed by the great French mathematician Charles Hermite in the following words: 'Il existe, si je ne me trompe, tout un monde qui est l'ensemble des vérités mathématiques, dans lequel nous n'avons d'accès que par l'intelligence, comme existe le monde des réalités physiques; l'un et l'autre indépendant de nous, tous deux de création divine et qui ne semblent distincts qu'à cause de la faiblesse de notre esprit, par contre ne sont pour une pensée puissante qu'une seule et même chose, et dont la synthèse se rélève partiellement dans cette merveilleuse correspondence entre les Mathématiques abstraites d'une part, l'Astronomie, et toutes les branches de la Physique de l'autre'.

Only after mathematics had been recognized as an autonomous interior constructional activity which, although it can be *applied to* an exterior world, neither in its origin nor in its methods *depends on* an exterior world, firstly all axioms became illusory, and secondly the criterion of truth or falsehood of a mathematical assertion was confined to mathematical activity itself, without appeal to logic or to hypothetical omniscient beings. An immediate consequence was that for a mathematical assertion α the two cases of truth and falsehood, formerly exclusively admitted, were replaced by the following three:

(1) α has been proved to be true;

(2) α has been proved to be absurd;

(3) α has neither been proved to be true nor to be absurd, nor do we know a finite algorithm leading to the statement either that α is true or that α is absurd.†

In contrast to the perpetual character of cases (1) and (2), an assertion of type (3) may at some time pass into another case, not only because further thinking may generate an algorithm accomplishing this passage, but also because in modern or *intuitionistic* mathematics, as we shall see presently, a mathematical entity is not necessarily predeterminate, and may, in its state of free growth, at some time acquire a property which it did not possess before. [Next follows the introduction of fleeing properties and the familiar counterexamples, cf. p. 6ff, or Brouwer (1955).]

† The case that α has neither been proved to be true nor to be absurd, but that we know a finite algorithm leading to the statement either that α is true, or that α is absurd, obviously is reducible to the first and second cases. This applies in particular to assertions of possibility of a construction of bounded finite character in a finite mathematical system, because such a construction can be attempted only in a finite number of particular ways, and each attempt proves successful or abortive in a finite number of steps.

One of the reasons† that led intuitionistic mathematics to this extension was the failure of classical mathematics to compose the continuum out of points without the help of logic. For, of real numbers determined by predeterminate convergent infinite sequences of rational numbers, only an *ever-unfinished denumerable* species can actually be generated. This ever-unfinished denumerable species being condemned never to exceed the measure zero, classical mathematics, in order to compose a continuum of positive measure out of points, has recourse to some logical process starting from at least an axiom. A rather common method of this kind is due to Hilbert who, starting from a set of properties of order and calculation, including the Archimedean property, holding for the arithmetic of the field of rational numbers, and considering successive extensions of this field and arithmetic to the extended fields and arithmetics conserving the foresaid properties, including the preceding fields and arithmetics, postulates the existence of an ultimate such extended field and arithmetic incapable of further extension, i.e. he applies the so-called axiom of completeness. From the intuitionistic point of view the continuum created in this way has a merely linguistic, and no mathematical, existence. It is only by means of the admission of freely proceeding infinite sequences that intuitionistic mathematics has succeeded to replace this linguistic continuum by a genuine mathematical continuum of positive measure, and the linguistic truths of classical analysis by genuine mathematical truths.

However, notwithstanding its rejection of classical logic as an instrument to discover mathematical truths, intuitionistic mathematics has its general introspective theory of mathematical assertions. This theory, which with some right may be called intuitionistic mathematical logic, we shall illustrate by the following remarks. [Now a paragraph follows of roughly the same

† Added in the margin of the manuscript: Incorrect, the extension is an immediate consequence of the selfunfolding; so here only the *utility* of the extension is explained.]

content as p. 10ff. The following lines precede a number of counterexamples.]

Theorems holding in intuitionistic, but not in classical, mathematics often originate from the circumstance that for mathematical entities belonging to a certain species the inculcation of a certain property imposes a special character on their way of development from the basic intuition; and that from this compulsory special character properties ensue which for classical mathematics are false. Striking examples are the modern theorems that the *continuum does not split*, and that *a full function of the unit continuum is necessarily uniformly continuous.*

Notes

1. Ever since his inaugural address 'Intuitionism and Formalism' (Brouwer, 1912), Brouwer has opened his larger expository talks with historical remarks. In this address Brouwer uses the terms *formalism* and *intuitionism* for Hilbert's and Peano's old formalism and for the French pre-intuitionism. Some authors (e.g. Fraenkel) have used the term neo-intuitionism for Brouwer's brand of intuitionism.
2. Brouwer did not go beyond this cautious and rather dated appraisal of the use of logic. His reluctance to accept the conveniences of modern logic made him express himself here in colourful but rather uninformative terms. It may safely be ventured that the inferences of intuitionistic logic qualify equally well as the principles of contradiction and syllogism.
3. The first use of undecidable properties of effectively presented objects (such as the decimal expansion of π) occurs in Brouwer (1908). Fleeing properties are introduced in Brouwer (1929).
4. Brouwer's introduction of κ_f, the existence of which is still in doubt, is a perfect illustration of a *description* as presented by Dana Scott (1979) in the context of partial elements.
5. Fundamental sequences are assumed to be 'predeterminate' ('prädestiniert' in Brouwer (1930)), or 'lawlike' as we would say now.
6. By 'infinite decimal fractions having no exact value' Brouwer clearly means 'decimal fractions not given by a law'.
7. For convenience we repeat the definitions of various notions in a more compact notation:
   (i) a and b are *different*: $a \neq b$ iff $a = b \to \bot$;
   (ii) For $\alpha, \beta \in S^N$: $\alpha = \beta$ iff $\forall n(\alpha(n) = \beta(n))$;
   (iii) S is *discrete*: $\forall x \in S \forall y \in S(x = y \vee x \neq y)$;
   (iv) M *deviates* from N: $\exists x \in M \; \forall y \in N(x \neq y)$;
   (v) M is a *subspecies* of N (notation $M \subset N$): $\forall x(x \in M \to x \in N)$;
   (vi) M is a *proper subspecies* of N: $M \subset N \wedge M$ deviates from N;
   (vii) M is a *removable subspecies* of N: $M \subset N \wedge \forall x \in N(x \in M \vee x \notin M)$;
   (viii) $M = N$, M is equal to N: $\forall x \in M \; \exists y \in N(x = y) \wedge \forall y \in N \; \exists x \in M(x = y)$;
   (ix) M is *congruent* with N: $\neg \exists x \in M \forall y \in N(x \neq y) \wedge \neg \exists y \in N \forall x \in M(x \neq y)$.
8. It is tacitly understood that the species S in $\cap S$ contains an element. This is in the tradition of the early texts on set theory, e.g. those of Schoenfliess or Hausdorff. One might think that, since species are built from previously constructed elements, there is no harm in taking an empty intersection.

However, there is no bound on the universe of all species so the empty intersection is, in general, inadmissible.

9. Brouwer used $\mathscr{D}(S)$ and $\mathscr{P}(S)$, as in Brouwer (1918).
10. $\mathscr{D}(s_1, s_2, \ldots, s_n)$, $\mathscr{P}(s_1, \ldots, s_n)$, $\mathscr{D}(s_1, s_2, s_3, \ldots)$, $\mathscr{P}(s_1, s_2, s_3, \ldots)$ are used in the manuscript.
11. The theorem $\neg\neg\neg A \Leftrightarrow \neg A$ was established in Brouwer (1923).
12. The notion of *barrage* is new. It is so weak that Brouwer has not made any use of it. The central notion here is that of *crude block*, cf. Brouwer (1954, p. 9).

    We continue our transcription of Brouwer's definitions:

    (i) $K(\sigma) = \{n \in \mathrm{SEQ} | \forall m \prec n \ \neg\mathrm{St}(m)\}$;

    (ii) $\mathrm{FA} = \{\xi | \forall x \neg\mathrm{St}(\bar{\xi}x)\}$, the species of free arrows;

    (iii) $\mathrm{SA} = \{f | \forall x \neg\mathrm{St}(\bar{f}x)\}$, the species of sharp free arrows;

    (iv) $\beta(\sigma)$ is a *barrage* $\Leftrightarrow \neg\exists\xi \in \mathrm{FA} \ \forall x(\bar{\xi}x \notin \beta(\sigma))$, where ξ varies over free arrows (choice sequences);

    (v) $\beta(\sigma)$ is a *crude block* $\Leftrightarrow \forall\xi \in \mathrm{FA} \ \exists x(\bar{\xi}x \in \beta(\sigma))$;

    (vi) a *massive crude block* is a crude block which is hereditary under descendants;

    (vii) $B(\sigma)$ is a (proper) *block* $\Leftrightarrow \forall\xi \in \mathrm{FA} \ \exists x(\bar{\xi}x \in B(\sigma)) \wedge \forall n(n \in B(\sigma) \vee n \notin B(\sigma)) \wedge \forall n \prec m(m \in B(\sigma) \rightarrow n \notin B(\sigma))$;

    (viii) the *stump* $B''(\sigma)$ carried by the block $B(\sigma)$ is given by $B''(\sigma) = \{n | \neg\exists m \in B(\sigma)(m \prec n)\}$.

    Geometrically, a stump carried by a block is an upper part of a tree with nodes of the block at the bottom.

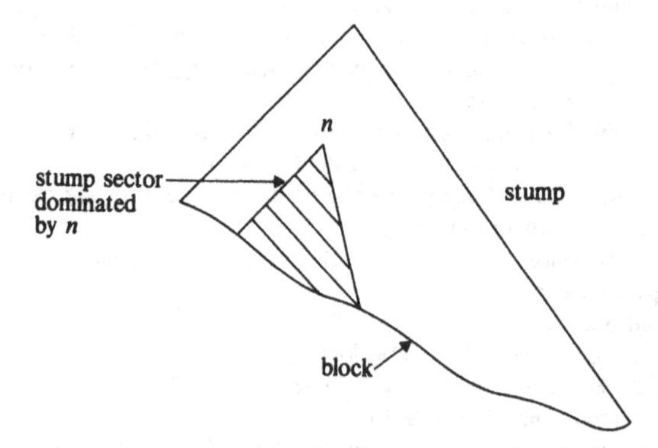

13. Brouwer introduces a number of properties of spread directions which yield a refined classification. It is not always clear why the notions were introduced, since in most applications only the more basic kinds of spreads turn up.

    The species σ is of *clear opening* if the non-sterilized nodes $\langle x \rangle$ of length 1 are 'strongly infinite' or 'bounded', i.e. $\forall x \exists y > x \neg\mathrm{St}(\langle y \rangle) \vee \exists z \forall w > z \ \mathrm{St}(\langle w \rangle)$, abbreviated $\mathrm{Si}_0 \wedge \mathrm{Bo}_0$, or Cl_0. If all the non-sterilized nodes $\langle x \rangle$ precede the sterilized ones, $\forall xy(\mathrm{St}(\langle x \rangle) \wedge x < y \rightarrow \mathrm{St}(\langle y \rangle))$, abbreviated Cnd_0, σ is said to

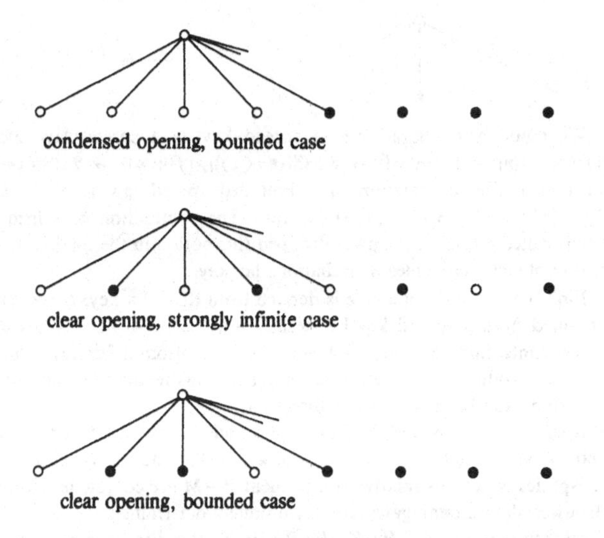

be of *condensed opening*. If $\exists z \forall w > z$ St($\langle w \rangle$), then σ is of *limited opening*.

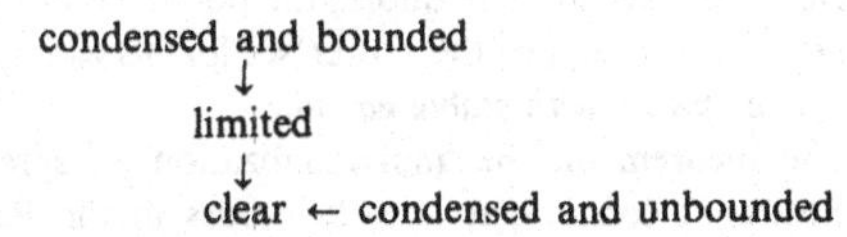

The notions of '...*continuation*' are completely similar. Instead of nodes $\langle x \rangle$ consider nodes $n * \langle x \rangle$ for a fixed node n ($n * m$ is the concatenation of n and m). We use the analogous abbreviations $\mathrm{Cl}_n = \mathrm{Si}_n \vee \mathrm{Bo}_n$, Cnd_n.

By imposing predeterminateness on various properties Brouwer further specializes the notions of spread direction. We will reproduce those notions here in a more formal notation. A spread direction σ is said to be:

(i) *clear cut* if $\exists f, g \forall n[\neg \mathrm{St}(n) \rightarrow [(f(n)=0 \rightarrow \mathrm{Si}_n) \wedge (f(n) \neq 0 \rightarrow \forall w > g(n) \mathrm{St}(n * \langle w \rangle))]]$;

(ii) *solid* if it is clear cut and $\forall n(\neg \mathrm{St}(n) \rightarrow \mathrm{Cnd}_n)$ (since it is possible to decide whether a node is sterilized or not we can replace the g from (i) by g' with $g'(n) = \mu x \; \mathrm{St}(n * \langle x \rangle)$);

(iii) *bounded* if $\exists g \forall n(\neg \mathrm{St}(n) \rightarrow \forall w > g(n) \mathrm{St}(n * \langle w \rangle))$;

(iv) *firm* if it is solid and bounded, which can also be expressed as $\exists g \; \forall n[\neg \mathrm{St}(n) \rightarrow \forall w((w < g(n) \rightarrow \neg \mathrm{St}(n * \langle w \rangle)) \wedge (w > g(n) \rightarrow \mathrm{St}(n * \langle w \rangle)))]$.

14. The passages between open brackets **[ ]** have been struck out in the final version of the manuscript. They are reproduced here because the notions introduced in them have none the less been used in the sequel.

15. A spread key is obtained here by a kind of contraction from a clear-cut spread law: remove the sterilized nodes and shift the non-sterilized ones to the left.

We could also characterize a spread key as a spread law such that $\exists f \forall n(\neg \mathrm{St}(n) \rightarrow [(f(n)=0 \rightarrow \forall x \neg \mathrm{St}(n * \langle x\rangle)) \wedge (f(n) \neq 0 \rightarrow \forall x(\mathrm{St}(n * \langle x\rangle \leftrightarrow x>f(n))])$. The contraction of a bounded spread law is a *fan key*, i.e. $\exists f \forall n(\neg \mathrm{St}(n) \rightarrow \forall x(\neg \mathrm{St}(n * \langle x\rangle \leftrightarrow x<f(n)))$. The contraction of a firm spread law is called a *cluster*. Brouwer dropped this notion in the final version; as a matter of fact it coincides with that of a fan key.

Finally the notion of a *clue* is derived from that of a key. A *spread clue* is obtained from a spread key by sterilizing a species of nodes, closed under descendants. Similarly one obtains a *bunch clue* from a fan key. The reason for the introduction of these clues is not clear as it takes us back to spread directions and bounded spread directions.

16. P is *composed* of M and N if $M \cap N=\emptyset$ and $\neg \exists x \in P(x \notin M \wedge x \notin N)$. Such M and N are called *conjugate subspecies* of P. One easily checks that a subspecies M and its relative complement $P-M$ are conjugate. However, as Brouwer shows, conjugates are not uniquely determined.
17. P *splits* into H and K if $H \cap K=\emptyset \wedge P=H \cup K$. P is *directly composed* of V and W if $\forall x \in P((x \in V \leftrightarrow x \notin W) \wedge(x \in W \leftrightarrow x \notin V))$.
18. Here Brouwer literally quotes the power set of the reals. Note, however, that he does not require the power set for his example.
19. *P is a species with stable equality.*
20. The theorem on the 'individualization' of spreads figures prominently in Brouwer's lectures, it already occurs in the Berlin lectures and has been published in Brouwer (1928). It states that for any spread (with decidable equality on its species of figures) there can be constructed a subspecies of a spread with a one-to-one assignment of sequences of figures to nodes.
21. As the diagram does not determine the spread law, let us add that the spread assigns the last index to the initial segments in the diagram and sterilizes all other nodes.
22. Here Brouwer introduces Cartesian spaces of finite dimension by means of nested intervals. Note that there is no prescribed modulus of convergence. This method goes back to Brouwer (1919). Most of the topological notions (including some refinements) are also to be found in that publication.
23. Brouwer consistently considers, in his treatment of Cartesian space, the case of the reduced space separately. In modern terminology one would speak of the lawlike space, or the lawlike continuum. Thus Brouwer's position on the subject of lawlike functions seems to be closer to that of Kreisel than to that of Kleene, in the sense that there is a good case for introducing 'lawlikeness' as a primitive notion in intuitionistic analysis.
24. Located point species were introduced in Brouwer (1919). The simplest way to recognize that each $k^{(\nu)}$ is superfluous or acceptable is to use the metric $d(x,y)=\max(|x_1-y_1|,|x_2-y_2|)$, under which the basic neighbourhoods are squares (cf. Heyting (1956, 5.2.1)).

25. In the case where we deal with a species such as the unity square, the resulting bunch is even a cluster, i.e. the bound on the number of acceptable next squares is given by a lawlike function, and there is a lawlike function that specifies the acceptable squares. A possible motive for introducing the various refinements of spreads could be the wish to consider finer distinctions in the classification of point spreads. Consider, for example, a singleton point species, i.e. $\{p\}$. If p is given by a sharp arrow, i.e. a lawlike function, then $\{p\}$ coincides with a cluster, while in general a singleton species will coincide with a bunch.
26. The general notion of 'located sequence' and 'located compact topological space' is introduced in Brouwer (1926b). The reader familiar with general topology may think of an enumeration of the ε-nets for $\varepsilon = 1, 2^{-1}, 2^{-2}, \ldots$ of a compact metric space.
27. Where $s_{\mu_n} = \{p_1, \ldots, p_{\mu_n}\}$.
28. In this counterexample (and in later ones) Brouwer restricts himself, in the case of the reduced (lawlike) continuum, to numbers (points) defined by reference to a fleeing property. As a result the counterexamples are weak, i.e. it is shown that we do not have, nor can expect, a proof of the statement under consideration. If one admits the sequences depending on the creative subject (empirical sequences) as lawlike, then one can expect stronger results. In this particular example one can choose concrete simple species $\beta(A)$ and $\gamma(A)$, e.g. of the even and odd numbers. The use of a variable pair makes sense if one is aiming for stronger counterexamples, as, for example, in Brouwer (1949a).
29. The statement 'neither <0, nor >0' has to be interpreted in the weak sense, i.e. 'it is an open problem whether $\pi_\alpha < 0$ or $\pi_\alpha > 0$'. The strong reading would lead to a contradiction.

    The remark in pencil clearly refers to the improvement in Brouwer (1949a), where it is shown that $\neg\forall x(x<0 \to \bar{x}<0)$ (or, in Brouwer (1949b), $\neg\forall x(x \neq 0 \to x<0 \vee x>0)$).
30. Well-ordering is introduced in Brouwer (1918). The early expositions are far more elaborate than the later ones, in which only the essentials required for the bar theorem are presented. The inductive definition employed by Brouwer is modelled after Cantor's first definition of ordinal and well-ordering in 1883.
31. Recall that a fundamental sequence is lawlike, therefore the well-ordered species are lawlike in nature.
32. Here Brouwer establishes the connection between well-founded trees and (countable) ordinals. He was one of the first, if not *the* first, to note and exploit this connection. The use of constructional subspecies goes back to Brouwer (1918) and earlier unpublished lecture notes.
33. The remark on the pseudo well-ordered structure of intuitionistic deductions appears for the first time in a footnote in Brouwer (1927a). Similar ideas were stated by Zermelo (1935).
34. The equivalence of inextensible and virtual order was first shown in Brouwer (1927b). In Brouwer (1975, p. 596), a note of Brouwer is reproduced in

which the definition of virtual order is reconsidered. The present elucidation is of the same purport.

The notion of virtual order is apparently in need of further explanation since the obvious traditional interpretation conflicts with the theorem on the equivalence of virtual and inextensible order.

Consider the (intuitionistic) structure $\langle\{a,b,c,d\}, <\rangle$ (Brouwer, 1975, p. 596), with $a<b$, $a<c$, $a<d$, $b<d$, $c<d$ and take for $\doteq$ the real identity. The properties (1)–(7) (p. 40) obviously hold, but properties (8) and (9) (p. 51) present a more delicate problem. Classically, one would say $b=d$ and $b<d$ do not hold in the structure, so that $b\neq d$ and $\neg b<d$ hold, hence by (8) $b>d$ should hold. However, intuitionistically $b\neq d$ cannot be claimed, as this would require enough information to reduce $b=d$ *ad absurdum*. So, intuitionistically, (8) and (9) are vacuously true, and hence the ordering is virtual. But, as Brouwer observes, the order can consistently be extended, e.g. by adding $b<c$.

In order to clear up these difficulties Brouwer added the clauses involving the 'possibilities to reduce relations from the criteria'. There are two ways in which one can implement Brouwer's remarks. The first consists of treating the whole problem as an exercise in proof theory, with '$\vdash$' for 'deducibility'; this is certainly not in the spirit of Brouwer, and is therefore correctly criticized by Heyting. The second, which was elegantly worked out by Posy, is obtained by interpreting 'deducible' as in the theory of the creative subject.

In this analysis we can, for example, write for (8) $\neg\exists x\vdash_x A=B\wedge\neg\exists x\vdash_x A<B\rightarrow\exists x\vdash_x A>B$ (by the axioms of the creative subject we can replace this by $\neg\exists x\vdash_x A=B\wedge\neg\exists x\vdash_x A<B\rightarrow A>B$). Reading '$\neg\exists x\vdash_x\neg X'<Y$'' for 'the relation $X'<Y'$ can be added in a non-contradictory way', one can now easily formalize Brouwer's argument.

35. The continuum, in the intuitionistic setting, has a much more refined and interesting structure than it has in ordinary classical mathematics. Brouwer has stressed this point ever since his first systematic expositions (cf. Brouwer, 1919). It was a topic of the Berlin lectures, and 'Die Struktur des Kontinuums' (Brouwer, 1930), the Vienna lecture of 1928, was exclusively devoted to an analysis of the continuum.

In the present chapter Brouwer returns to the continuum, but this time he bases the investigation on the notion of order. It seems natural to use some intuitionistic version of Dedekind cuts as, for example, used in Troelstra (1980). Brouwer instead, opts for a, prima facie, more constructive approximation procedure, introduced in Brouwer (1926a).

We will compare Brouwer's intercalation subdivision with the following version of Dedekind cut. A Dedekind cut is a pair $\langle L,R\rangle$ of subspecies of Q such that

(i) $\exists p\in L$, $\exists q\in R$, $L\cap R=\emptyset$;
(ii) $p\in L\rightarrow\exists q\in L(p<q)$, $q\in R\rightarrow\exists p\in R(p<q)$;
(iii) $p<q\rightarrow p\in L\vee q\in R$.

An intercalation subdivision determines a Dedekind cut as follows: Let $(g_i)_{i\geq 1}$ enumerate $\bar{\eta}(\varphi)$, define L_i as the left subspecies of φ_i, define R_i as the right subspecies of φ_i, and $L = \text{Interior} \cup L_i$, $R = \text{Interior} \cup R_i$; (i) and (ii) are then trivially satisfied. For (iii) consider $p = g_i, q = g_j$ and $i < j$. In φ_j at most one of g_i, g_j can be 'undetermined', i.e. the other one belongs to L_j or R_j. Hence (iii) follows.

Conversely, let a Dedekind cut $\langle L,R \rangle$ be given, and let (g_i) be an enumeration of $\bar{\eta}(\varphi)$. Put $\varphi_1 = \{g_1\}$ and $g_1 \in L_1 \Leftrightarrow g_1 \in L$, $g_1 \in R_1 \Leftrightarrow g_1 \in R$. Thus if $g_i \in L \cup R$ is not settled we leave g_1 undetermined in φ_1. Suppose φ_n has been defined, with g_i $(i \leqq n)$ as an undetermined element. Since $g_{n+1} \neq g_i$ we have $g_{n+1} < g_i$ or $g_i < g_{n+1}$. By (iii) we find that one of the elements g_i, g_{n+1} belongs to $L \cup R$, so we form φ_{n+1} accordingly. The *exhaustive sectional subdivision* of p. 71 is a variation on the Dedekind cut as presented above.

36. The Dedekind cuts mentioned here in connection with the elements of first-order precision position satisfy the additional (classical) condition $L \cup R = Q$.
37. The equivalence of 'having first-order precision with respect to the decimal fractions' and 'having a decimal expansion' is observed in Brouwer (1921).
38. The properties (1)–(5) correspond to the properties (3)–(7) of Brouwer (1930). Brouwer consistently distinguishes the cases of the reduced and the full continuum, probably with stronger results in mind for the latter case, since he uses the creative subject in the counterexamples. However, he does not use the full power, which would yield, for example, the contradictority of ordinal density in itself of the full continuum. Note that usually the counterexample for the reduced continuum already yields a counterexample for the full continuum.
39. This chapter presents one of the highlights of intuitionism. It contains a breakthrough in constructive mathematics, which lifts it successfully above the discrete arithmetic practice.

    Already in 1924 Brouwer established the general theorem, which in the language of trees can be roughly formulated as 'well-founded trees can be inductively defined'; a precise formulation can be found in Kleene & Vesley (1965), or Troelstra (1977). However, it was not the theorem itself that attracted attention, but rather a special form, restricted to *fans* (or compact trees), and its corollary, the uniform continuity theorem for functions on closed intervals.

    The applications to analysis and topology that Brouwer made required only the fan theorem. However, the bar theorem (here 'every crude spread block contains a well-ordered block') is an outstanding example of a truly conceptual meta-mathematical result.

    It is obtained by reflection on the possible proof of a statement of the form $\forall \xi \exists x\, A(\xi x)$. At this place Brouwer essentially uses the fact that mathematical derivations are in the form of (pseudo) well-ordered species. The well-ordering of such a derivation is used to establish, by transfinite induction, the well-ordering property of a suitable block contained in the given block. Brouwer's 'no arrow containing this node can avoid *B*' must be a slip of the pen. For by definition 'every arrow containing this node must meet *B*'.

    Brouwer's aim of reducing the principle of bar induction to the basic intuitionistic notion of 'deduction' cannot be considered to be achieved.

Kleene has shown that the unrestricted principle of bar induction conflicts with the continuity principles (cf. Brouwer, 1975, p. 609).

We will reproduce an informal simplified version. Goldbach's conjecture reads 'each even natural number is the sum of two odd primes'; let us abbreviate '$2n$ is the sum of two odd primes' as $G(n)$. Now consider the following species B, defined by $(\langle\rangle \in B \leftrightarrow \neg\forall x\, G(x)) \wedge (\langle n\rangle \in B \leftrightarrow G(n))$. It is clear that B is a crude block. However, B does not contain a well-ordered block; suppose that B were to contain the well-ordered block B', then $\langle\rangle \in B \vee \langle\rangle \notin B$. Hence it would follow that $\neg\forall x\, G(x) \vee \forall x\, G(x)$, which has not been established (in general we would obtain the decidability of all Π_1^0 statements).

We must conclude that Brouwer's conceptual basis for the bar theorem is insufficient. In particular his analysis of the elementary steps (ζ- and F-inference) are questionable. For an extensive analysis the reader is referred to Dummett (1977, § 3.4).

40. This application of the non-splittability has not appeared earlier, but there is an undated note, which by the handwriting seems to belong to the thirties, that contains essentially the same material.

References

Brouwer, L. E. J. (1908) De onbetrouwbaarheid der logische principes. *Tijdschrift voor Wijsbegeerte* **2**, 152–8. Translated as 'The unreliability of the logical principles' in Brouwer (1975), pp. 107–11.

– (1912) Intuitionism and formalism (translated by A. Dresden). *Bulletin of the American Mathematical Society* **20** (1913), 81–96.

– (1918) Begründung der Mengenlehre unabhängig vom logischen Satz vom ausgeschlossenen Dritten. Erster Teil, Allgemeine Mengenlehre. *Koninklijke Nederlandse Akademie van Wetenschappen, Verhandelingen*, 1e sectie, **12**, no. 5, 43 pp.

– (1919) Begründung der Mengenlehre unabhängig vom logischen Satz vom ausgeschlossenen Dritten. Zweiter Teil, Theorie der Punktmengen. *Koninklijke Nederlandse Akademie van Wetenschappen, Verhandelingen*, 1e sectie, **12**, no. 7, 33 pp.

– (1921) Besitzt jede reelle Zahl eine Dezimalbruch-Entwickelung? *Koninklijke Akademie van Wetenschappen te Amsterdam* (Verslag van de gewone vergaderingen der wis- en natuurkunde afdeling) **29**, 803–12. Also: *Koninklijke Nederlandse Akademie van Wetenschappen, Proceedings* **23**, 955–64. Also: *Mathematische Annalen* **83**, 201–10.

– (1923) Intuitionistische Zerlegung mathematischer Grundbegriffe. *Jahresbericht der Deutschen Mathematiker-Vereinigung* **33** (1925), 251–6.

– (1926a) Zur Begründung der intuitionistischen Mathematik II. *Mathematische Annalen* **95**, 453–72.

– (1926b) Intuitionistische Einführung des Dimensionsbegriffes. *Koninklijke Nederlandse Akademie van Wetenschappen, Proceedings* **29**, 855–63.

– (1927a) Über Definitionsbereiche von Funktionen. *Mathematische Annalen* **97**, 60–75. (English translation of §§1–3: On the domains of definition of functions, in J. van Heyenoort, *From Frege to Gödel*, Cambridge, Mass. 1967, 446–63.)

– (1927b) Virtuelle Ordnung und unerweiterbare Ordnung. *Journal für die reine und Angewandte Mathematik* **157**, 255–7.

– (1928) Beweis, dass jede Menge in einer individualisierten Menge enthalten ist. *Koninklijke Akademie van Wetenschappen, Proceedings* **31**, 380–1.

– (1929) Mathematik, Wissenschaft und Sprache. *Monatshefte für Mathematik und Physik* **36**, 153–64.

– (1930) *Die Struktur des Kontinuums.* Gistel, Vienna, 14 pp.

– (1942b) Beweis dass der Begriff der Menge höherer Ordnung nicht als Grundbegriff der intuitionistischen Mathematik in Betracht kommt. *Indagationes Mathematicae* **4**, 274–6.

– (1949a) De non-aequivalentie van de constructieve en de negatieve orderelatie in het continuum. *Indagationes Mathematicae* **11**, 37–9. Translated as 'The non-equivalence of the constructive and the negative order relation on the continuum' in Brouwer (1975, pp. 495–7).

– (1949b) 'Contradictoriteit der elementaire meetkunde. *Indagationes Mathematicae* **11**, 89–90. Translated as 'Contradictority of elementary geometry' in Brouwer (1975, pp. 497–8).

– (1952) Historical background, principles and methods of intuitionism. *South African Journal of Science* **49**, 139–46.

– (1954) Points and spaces. *Canadian Journal of Mathematics* **6**, 1–17.

– (1955) The effect of intuitionism on classical algebra of logic. *Proceedings of the Royal Irish Academy*, Section A, **57**, 113–16.

– (1975) *Collected Works*, Vol. I, ed. A. Heyting. North-Holland Publ. Co., Amsterdam. 628 pp.

van Dalen, D. (1978) Brouwer: The Genesis of his Intuitionism. *Dialectica* **32**, 291–303.

Dummett, M. (1977) *Elements of Intuitionism.* Oxford University Press, 480 pp.

Heyting, A. (1956) *Intuitionism: An Introduction.* North-Holland Publ. Co., Amsterdam. 145 pp.

Kleene, S. C. & Vesley, R. E. (1965) *The Foundations of Intuitionistic Mathematics.* North-Holland Publ. Co., Amsterdam. 206 pp.

Posy, C. J. (1980) A note on Brouwer's definition of unextendable order. *History and Philosophy of Logic* **1**, 139–49.

Scott, D. S. (1979) Identity and existence. In *Applications of sheaves*, ed. M. P. Fourman, C. J. Mulvey & D. S. Scott, pp. 660–96. Springer-Verlag, Berlin.

Troelstra, A. (1969) *Principles of intuitionism.* Lecture Notes in Mathematics no. 95, Springer-Verlag, Berlin. 111 pp.

– (1977) *Choice Sequences. A Chapter of Intuitionistic Mathematics.* Oxford University Press. 180 pp.

– (1980) Intuitionistic extensions of the reals. *Nieuw Archief voor Wiskunde* **28**, 63–113.

Zermelo, E. (1935) Grundlagen einer allgemeinen Theorie der mathematischen Satzsystemen. *Fundamenta Mathematicae* **25**, 136–46.

Index